Maintenance Fundamentals

Maintenance Fundamentals

Editor

Ravindra Kumar

Maintenance Fundamentals
Edited by **Ravindra Kumar**

Printed in 2017

ISBN: 978-1-68117-410-5

Library of Congress Control Number: 2015941617

© 2016 by
SCITUS Academics LLC,
616, Corporate Way, Suite 2, 4766,
Valley Cottage, NY 10989

www.scitusacademics.com

Notice

Reasonable efforts have been made to publish reliable data and views articulated in the chapters are those of the individual contributors, and not necessarily those of the editors or publishers. Editors or publishers are not responsible for the accuracy of the information in the published chapters or consequences of their use. The publisher believes no responsibility for any damage or grievance to the persons or property arising out of the use of any materials, instructions, methods or thoughts in the book. The editors and the publisher have attempted to trace the copyright holders of all material reproduced in this publication and apologize to copyright holders if permission has not been obtained. If any copyright holder has not been acknowledged, please write to us so we may rectify.

Contents

Preface

Maintenance Fundamentals offers a full account of thermodynamic systems in chemical engineering. It provides a solid understanding of the basic concepts of the laws of thermodynamics as well as their applications with a thorough discussion of phase and chemical reaction equilibrium. At the outset the text explains the various key terms of thermodynamics with suitable examples and then thoroughly deals with the virial and cubic equations of state by showing the P-V-T (pressure, molar volume and temperature) relation of fluids. It elaborates on the first and second laws of thermodynamics and their applications with the help of numerous engineering examples. The text further discusses the concepts of energy, standard property changes of chemical reactions, thermodynamic property relations and fugacity. The book also includes detailed discussions on residual and excess properties of mixtures, various activity coefficient models, local composition models, and group contribution methods. In addition, the text focuses on vapour-liquid and other phase equilibrium calculations, and analyses chemical reaction equilibrium and adiabatic reaction temperature for systems with complete and incomplete conversion of reactants. The book is primarily designed for the undergraduate students of chemical engineering and its related disciplines such as petroleum engineering and polymer engineering. It can also be useful to professionals.

Editor

Chapter 1

An Overview of Fuel Cell Technology: Fundamentals and Applications

Omar Z. Sharaf and Mehmet F. Orhan

Department of Mechanical Engineering, College of Engineering, American University of Sharjah, Sharjah, United Arab Emirates

ABSTRACT

This paper provides a comprehensive review of fuel cell science and engineering with a focus on hydrogen fuel cells. The paper provides a concise, up-to-date review of fuel cell fundamentals; history; competing technologies; types; advantages and challenges; portable, stationary, and transportation applications and markets; current status of research-and-development; future targets; design levels; thermodynamic and electrochemical principles; system evaluation factors; and prospects and outlook. The most current data from industry and academia have been used with the relation between fuel cell fundamentals and applications highlighted throughout the manuscript.

INTRODUCTION

A fuel cell is an electrochemical device that converts the chemical energy of a fuel directly into electrical energy. The one-step (from chemical to electrical energy) nature of this process, in comparison to the multi-step (e.g. from chemical to thermal to mechanical to electrical energy) processes involved in combustion-based heat engines, offers several unique advantages. For instance, the current combustion-based energy generation technologies are very harmful to the environment and are predominantly contributing to many global concerns, such as climate change, ozone layer depletion, acidic rains, and thus, the consistent reduction in the vegetation cover. Furthermore, these technologies depend on the finite and dwindling world supplies of fossil fuels.

Fuel cells, on the other hand, provide an efficient and clean mechanism for energy conversion. Additionally, fuel cells are compatible with renewable sources and modern energy carriers (i.e., hydrogen) for sustainable development and energy security. As a result, they are regarded as the energy conversion devices of the future. The static nature of fuel cells also means quiet operation without noise or vibration, while their inherent modularity allows for simple construction and a diverse range of applications in portable, stationary, and transportation power generation. In short, fuel cells provide a cleaner, more efficient, and possibly the most flexible chemical-to-electrical energy conversion.

Polymer electrolyte membrane, also proton exchange membrane, fuel cells (PEMFC) in particular are one of the most promising types already in the early commercialization stage. Nonetheless, further development and research are required in order to reduce their costs, enhance their durability, and further optimize and improve their performance. Most of the research currently being conducted on PEMFCs is on the individual cell-level and the general system-level. Stack-level research, on the other hand, is an area that requires further research and development.

A proper understanding of the principles of fuel cell operation combined with a current outlook of the fuel cell industry are vital for overcoming current obstacles and the general advancement of fuel cell technology. Nevertheless, fuel cells are an interdisciplinary science in which electrochemistry, thermodynamics, engineering economics, material science and engineering, and electrical engineering all combine; making this a difficult task. This paper provides an up-to-date overlook of the fuel cell industry coupled with a concise digest of fuel cell operation principles as a contribution to the ongoing efforts to promote and commercialize fuel cells. We will often attempt to highlight the relations between a fuel cell's principals of operation, features and advantages, and areas of applications throughout the manuscript. This relation between principals, features, and applications is outlined in Fig. 1.

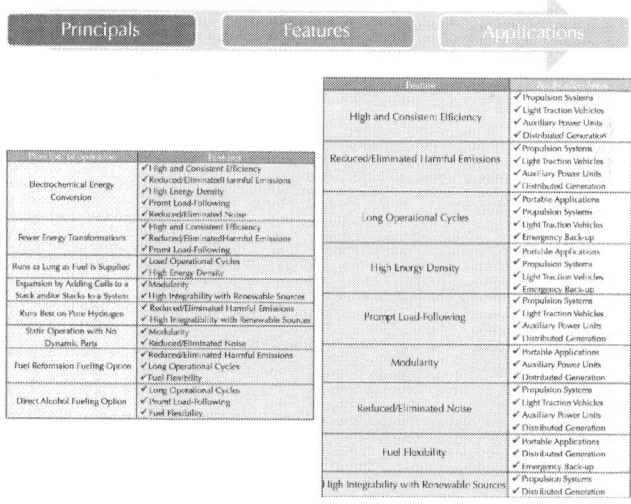

Figure 1: Outline of the relations between a fuel cell's principals of operation, advantages and features, and main areas of application.

AN OVERVIEW OF FUNDAMENTALS

A fuel cell is composed of three active components: a fuel electrode (anode), an oxidant electrode (cathode), and an electrolyte sandwiched between them. The electrodes consist of a porous material that is covered with a layer of catalyst (often platinum in PEMFCs). Fig. 2 illustrates the basic operational processes within a typical PEMFC [1]. Molecular hydrogen (H_2) is delivered from a gas-flow stream to the anode where it reacts electrochemically. The hydrogen is oxidized to produce hydrogen ions and electrons, as shown in Fig. 2, per the following equation:

$$H_2 \Rightarrow 2H^+ + 2e^- \qquad (1)$$

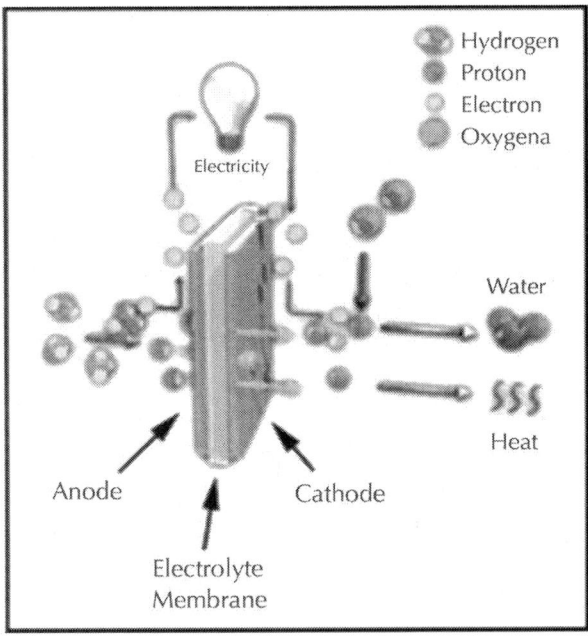

Figure 2: Typical PEM fuel cell operation [1].

The hydrogen ions migrate through the acidic electrolyte while the electrons are forced through an external circuit all the way to the cathode. At the cathode, the electrons and the hydrogen ions react with the oxygen supplied from an external gas-flow stream to form water, as shown in Fig. 2, per the following equation:

$$\frac{1}{2}O_2 + 2H^+ + 2e^- \Rightarrow H_2O$$

(2)

The overall reaction in the fuel cell produces water, heat, and electrical work as follows:

$$H_2 + \frac{1}{2}O_2 \Rightarrow H_2O + W_{ele} + Q_{heat}$$

(3)

The heat and water by-products must be continuously removed in order to maintain continuous isothermal operation for ideal electric power generation. Hence, water and thermal management are key areas in the efficient design and operation of fuel cells.

History

Research and development that eventually led to a functional fuel cell goes back to the early 1800s. Sir William Grove, a chemist and patent lawyer, is broadly considered to be the father of fuel cell science due to his famous water electrolyzer/fuel cell experimental demonstration. Sir William Grove used his background of electrolysis to conceptualize a reverse process that could be used to generate electricity. Based on this hypothesis, Grove succeeded in building a device that combines hydrogen and oxygen to produce electricity (instead of separating them using electricity). The device, originally labeled a gas battery, came to be known as a fuel cell. Further research continued into the twentieth century. In 1959, Francis Thomas Bacon, an English engineer, demonstrated the first fully-operational fuel cell. His work was impressive enough to get licensed and adopted by NASA [2]. PEMFCs and alkaline fuel cells (AFCs), in particular, were practically used by NASA in the 1960s as part of the Gemini and Apollo manned space programs. The NASA fuel cells were customized, non-commercial, experienced several malfunctions, and used pure oxygen and hydrogen as an oxidant and fuel, respectively. Fuel cells nowadays; however, are used in transportation, stationary, and portable applications; are gradually being adopted by the public and private sectors; are becoming more reliable and durable for long-term operation; and can function using air and reformation-based hydrogen as an oxidant and fuel, respectively. Table 1 highlights the main milestones in the history of fuel cells.

Table 1: Main milestones in the history of fuel cells

Year(s)	Milestone
1839	W.R. Grove and C.F. Schönbein separately demonstrate the principals of a hydrogen fuel cell
1889	L. Mond and C. Langer develop porous electrodes, identify carbon monoxide poisoning, and generate hydrogen from coal
1893	F.W. Ostwald describes the functions of different components and explains the fundamental electrochemistry of fuel cells
1896	W.W. Jacques builds the first fuel cell with a practical application
1933–1959	F.T. Bacon develops AFC technology
1937–1939	E. Baur and H. Preis develop SOFC technology
1950	Teflon is used with platinum/acid and carbon/alkaline fuel cells
1955–1958	T. Grubb and L. Niedrach develop PEMFC technology at General Electric
1958–1961	G.H.J. Broers and J.A.A. Ketelaar develop MCFC technology
1960	NASA uses AFC technology based on Bacon's work in its Apollo space program
1961	G.V. Elmore and H.A. Tanner experiment with and develop PAFC technology
1962–1966	The PEMFC developed by General Electric is used in NASA's Gemini space program
1968	DuPont introduces Nafion*
1992	Jet Propulsion Laboratory develops DMFC technology
1990s	Worldwide extensive research on all fuel cell types with a focus on PEMFCs
2000s	Early commercialization of fuel cells

Competing Technologies

Fig. 3 shows a very interesting comparison between the main energy conversion devices in the market today [3]. It shows the typical exergy efficiencies of photovoltaic panels, thermal solar power plants, waste incineration, gas turbines, diesel engines, gas engines, Rankin cycles, combined Rankin–Brayton cycles, nuclear power plants, wind turbines, hydroelectric plants, in addition to fuel cells. As evident from the figure, fuel cells have one of the highest exergy efficiencies among the competing energy conversion devices. Table 2, Table 3 and Table 4, technoeconomically compare fuel cells to their competing technologies in the portable, stationary, and transportation sectors, respectively. The comparisons reveal that with respect to competing technologies, fuel cells have an advantage in both gravimetric and volumetric energy densities in the portable sector; have high efficiencies and capacity factors in the stationary sector; and offer high efficiencies and fuel flexibilities in the transportation sector. As such, it becomes clear that bringing down the investment cost of fuel cells in order to relatively level the competition field with other power technologies is probably the most vital challenge fuel cells need to overcome in order to become economically-feasible power generation alternatives. In the remainder of this section, we will further discuss the differences and similarities between fuel cells and their two closest energy competitors; namely—heat engines and batteries.

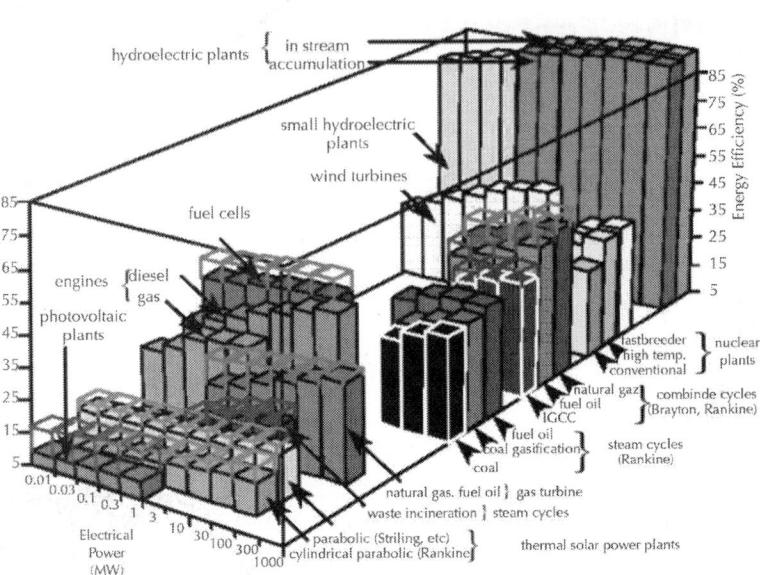

Figure 3: Exergy efficiencies of main energy conversion devices [3].

Table 2: Technoeconomic comparison between fuel cells and their competitors in the portable power sector (adapted from [5])

Portable power technology	Gravimetric energy density (Wh/kg)	Volumetric energy density (Wh/L)	Power density (W/kg)	Capital cost ($/kWh)
Direct methanol fuel cell	>1000	700–1000	100–200	200[a]
Lead-acid battery	20–50	50–100	150–300	70
Nickel–cadmium battery	40–60	75–150	150–200	300
Nickel–metal hydride battery	60–100	100–250	200–300	300–500
Lithium-ion battery	100–160	200–300	200–400	200–700
Flywheel	50–400	200	200–400	400–800
Ultracapacitor	10	10	500–10,000	20,000

[a]In $/kW.

Table 3: Technoeconomic comparison between fuel cells and their competitors in the stationary power/CHP sector (adapted from [5])

Stationary power/CHP technology	Power level (MW)	Efficiencya(%)	Lifetime (years)	Capital cost ($/kW)	Capacity factor (%)
Phosphoric acid fuel cell	0.2–10	30–45	5–20	1500	Up to 95
MCFC/gas turbine hybrid	0.1–100	55–65	5–20	1000	Up to 95
SOFC/gas turbine hybrid	0.1–100	55–65	5–20	1000	Up to 95
Steam cycle (coal)	10–1000	33–40	>20	1300–2000	60–90
Integrated gasification combined cycle	10–1000	43–47	>20	1500–2000	75–90
Gas turbine cycle (natural gas)	0.03–1000	30–40	>20	500–800	Up to 95
Combined gas turbine cycle (natural gas)	50–1000	45–60	>20	500–1000	Up to 95
Microturbine	0.01–0.5	15–30	5–10	800–1500	80–95
Nuclear	500–1400	32	>20	1500–2500	70–90
Hydroelectric	0.1–2000	65–90	>40	1500–3500	40–50
Wind turbine	0.1–10	20–50	20	1000–3000	20–40
Geothermal	1–200	5–20	>20	700–1500	Up to 95
Solar photovoltaic	0.001–1	10–15	15–25	2000–4000	<25

aFrom energy input to electrical output.

Table 4: Technoeconomic comparison between fuel cells and their competitors in the transportation propulsion sector (adapted from [5])

Transportation propulsion technology	Power level (kW)	Efficiency[a] (%)	Specific power (kW/kg)	Power density (kW/L)	Vehicle range (km)	Capital cost ($/kW)
Proton exchange membrane fuel cell (on-board fuel processing)	10–300	40–45	400–1000	600–2000	350–500	100
Proton exchange membrane fuel cell (off-board hydrogen)	10–300	50–55	400–1000	600–2000	200–300	100
Gasoline engine	10–300	15–25	>1000	>1000	600	20–50
Diesel engine	10–200	30–35	>1000	>1000	800	20–50
Diesel engine/battery hybrid	50–100	45	>1000	>1000	>800	50–80
Gasoline engine/battery hybrid	10–100	40–50	>1000	>1000	>800	50–80
Lead-acid or nickel-metal hydride battery	10–100	65	100–400	250–750	100–300	>100

[a]From energy input to electrical output.

Both fuel cells and heat engines typically use a hydrogen-based fluid and atmospheric air as the fuel and oxidant, respectively. However, fuel cells combine the fuel and oxidant electrochemically, while heat engines combine the fuel and oxidant via combustion. Additionally, fuel cells produce electrical work directly from chemical energy. While in the case of heat engines, producing electricity is a multi-step process that involves combustion to produce thermal energy from the internal chemical energy of the fuel. Then this thermal energy is converted into mechanical energy, and finally this mechanical energy is converted into electrical energy through the use of a generator. Generally, as the number of energy conversion processes increases in a certain device, the overall system efficiency of the device decreases. Fuel cells, in comparison to heat engines, produce fewer-to-zero pollutants and have higher theoretical and practical efficiencies. While on the other hand, heat engines are limited by the Carnot efficiency between their low and the high working temperatures and are responsible for a significant portion of the world's pollution. Finally, fuel cell stacks are static devices that operate with almost no noise or vibrations (as will be discussed in Section 3.5). Heat engines, on the other hand, have many dynamic components (e.g., pistons and gears) that produce a lot of noise and vibrations. This dynamic nature of heat engines limits their applications.

Fuel cells and batteries are quite similar in the sense that they are both electrochemical cells that consist of an electrolyte sandwiched between two electrodes. They both use internal oxidation–reduction reactions to convert the chemical energy content of a fuel to DC electricity. However, the composition and role of the electrodes differ significantly between the two energy devices. The electrodes in a battery are typically metals (e.g., zinc, lead, or lithium) immersed in mild acids. In fuel cells, the electrodes (i.e., catalyst layer and gas diffusion layer) typically consist of a proton-conducting media, carbon-supported catalyst, and electron-conducting fibers. Batteries are used as energy storage and conversion devices, while fuel cells are used for energy conversion only. A battery uses the chemical energy stored in its electrodes to fuel the electrochemical reactions that give us electricity at a specified potential difference. Thus, a battery has a limited lifetime and can only function as long as the electrodes' material is not yet depleted. Upon depletion of the electrodes' material, a battery must be either replaced (in case of a disposable battery) or recharged (by using an electric current to reform dissolved metals on the electrodes). In a fuel cell; however, the reactants are supplied from a separate storage device and the internal components are not used up in the electrochemical reactions. Thus, theoretically, a fuel cell can keep running as long as the reactants are sufficiently supplied and the products are properly removed. As a result, an operational fuel cell system requires a fuel storage mechanism and an oxidant supply mechanism to be incorporated within it. Moreover, when the battery is idle, electrochemical reactions that deteriorate the battery occur very slowly, reducing the lifetime of the battery. Rechargeable batteries suffer from many technical issues that limit their applications; some of these issues include power storage and retrieval potential, depth of charge, and number of charge/discharge cycles [4]. In the case of fuel cells this is not an issue. Additionally, there is no leakage or corrosion of cell components when the fuel cell is not in use, unlike with batteries [5]. Fig. 4[6] and Table 5 summarize the differences and similarities between fuel cells, heat engines, and batteries.

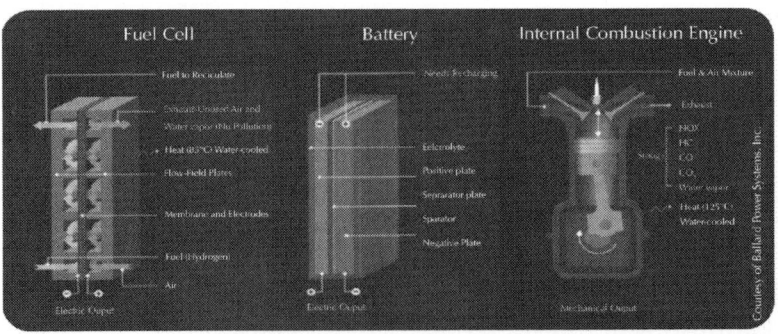

Figure 4: Fuel cell, battery, and internal combustion heat engine general structures [6].

Table 5: Summary of the similarities and differences between fuel cells, batteries, and heat engines

Comparison	Fuel cell	Battery	Heat engine
Function	Energy conversion	Energy storage & conversion	Energy conversion
Technology	Electrochemical reactions	Electrochemical reactions	Combustion
Typical fuel	Usually pure hydrogen	Stored chemicals	Gasoline, diesel
Useful output	DC electricity	DC electricity	Mechanical power
Main advantages	High efficiency	High efficiency	High maturity
	Reduced harmful emissions	High maturity	Low cost
Main disadvantages	High cost	Low operational cycles	Significant harmful emissions
	Low durability	Low durability	Low efficiency

Types

There are many types of fuel cells available in the market today. Fuel cells are conventionally categorized according to their electrolyte material. They differ in their power outputs, operating temperatures, electrical efficiencies, and typical applications. PEMFCs have the largest range of applications as they are extremely flexible. PEMFCs are the most promising candidates for transport applications due to their high power density, fast start-up time, high efficiency, low operating temperature, and easy and safe handling. However, PEMFCs are still too expensive to be competitive or economically-feasible. AFCs have the best performance when operating on pure hydrogen and oxygen, yet their intolerance to impurities (especially carbon oxides) and short lifetimes hinder their role for terrestrial applications (they are predominantly used for extraterrestrial purposes). Phosphoric acid fuel cells (PAFCs) are possibly the most commercially-developed fuel cells operating at intermediate temperatures.

Table 6: Fuel cell types according to electrolyte

Fuel cell type	Typical electrolyte	Typical anode/cathode catalysts[a]	Typical interconnect material	Typical fuel	Charge carrier[a]	Major contaminants[a]	Operation temperature (°C)	Specific advantages	Specific disadvantages	Electrical efficiency (%)	Technological maturity[b]	Research activity[c]
Low-temperature proton exchange membrane	Solid Nafion*	Anode: Platinum supported on carbon Cathode: Platinum supported on carbon	Graphite	Hydrogen	H^+	Carbon monoxide (CO) Hydrogen sulfide (H_2S)	60–80	Highly modular for most applications High power density Compact structure Rapid start-up due to low temperature operation Excellent dynamic response	Complex water and thermal management Low-grade heat High sensitivity to contaminants Expensive catalyst	40–60	4	H

	High-temperature proton exchange membrane
	Solid composite Nafion*
	Polybenzimidazole (PBI) doped in phosphoric acid
	Anode: Platinum–Ruthenium supported on carbon
	Cathode: Platinum–Ruthenium supported on carbon
	Graphite
	Hydrogen
	H+
	Carbon Monoxide (CO)
	110–180
	Simple water management
	Simple thermal management
	Accelerated reaction kinetics
	High-grade heat
	High tolerance to contaminants
	Accelerated stack degradation
	Humidification issues
	Expensive catalyst
	50–60
	3
	M

| Solid oxide | Solid yttria-stabilized zirconia (YSZ) | Anode: Nickel–YSZ composite Cathode: Strontium-doped lanthanum manganite (LSM) | Ceramics | Methane | O^{2-} | Sulfides | 800–1000 | High electrical efficiencies High-grade heat High tolerance to contaminants Possibility of internal reforming Eliminated electrolyte issues Fuel flexibility Inexpensive catalyst | Slow start-up Low power density Strict material requirements High thermal stresses Sealing issues Durability issues High manufacturing costs | 55–65 | 3 | H |

| Molten carbonate | Liquid alkali carbonate (Li_2CO_3, Na_2CO_3, K_2CO_3) in Lithium aluminate ($LiAlO_2$) | Anode: Nickel Chromium (NiCr) Cathode: Lithiated nickel oxide (NiO) | Stainless steel | Methane | CO_3^{2-} | Sulfides Halides | 600–700 | High electrical efficiencies; High-grade heat; High tolerance to contaminants; Possibility of internal reforming; Less strict material requirements; Fuel flexibility; Inexpensive catalyst | Slow start-up; Low power density; Electrolyte corrosion and evaporative losses; Corrosion of metallic parts; Air cross-over; Catalyst dissolution in electrolyte; Cathode carbon dioxide (CO_2) injection requirement | 55–65 | 4 | H |

| Phosphoric Acid | Concentrated liquid phosphoric acid (H_3PO_4) in silicon carbide (SiC) | Anode: Platinum supported on carbon

Cathode: Platinum supported on carbon | Graphite | Hydrogen | H^+ | Carbon monoxide (CO)

Siloxane

Hydrogen sulfide (H_2S) | 160–220 | Technologically mature and reliable

Simple water management

Good tolerance to contaminants

High-grade heat | Relatively slow start-up

Low power density

High sensitivity to contaminants

Expensive auxiliary systems

Low electrical efficiencies

Relatively large system size

Electrolyte acid loss

Expensive catalyst

High cost | 36–45 | 5 | M |

	Alkaline
Electrolyte	Potassium hydroxide (KOH) water solution; Anion exchange membrane (AEM)
Electrode	Anode: Nickel; Cathode: Silver supported on carbon
	Metallic wires
Fuel	Hydrogen
Charge carrier	OH⁻
Contaminant	Carbon dioxide (CO_2)
Temperature	Below zero–230
Advantages	High electric efficiency due to fast reduction reaction kinetics; Wide range of operation temperature and pressure; Inexpensive catalyst; Catalyst flexibility; Relatively low costs
Disadvantages	Extremely high sensitivity to contaminants; Pure hydrogen and oxygen required for operation; Low power density; Highly corrosive electrolyte leads to sealing issues; Complex and expensive electrolyte management for mobile electrolyte systems
	60–70
	5
	L

| Direct metha-nol | Solid Nafion® | Anode: Plati-num–Ru-thenium supported on carbon; Cathode: Platinum supported on carbon | Graphite | Liquid metha-nol–water solution | H⁺ | Carbon mon-oxide (CO) | Ambi-ent–110 | Compact size; Simple system; High fuel volumetric energy density; Easy fuel storage and delivery; Simple thermal manage-ment for liquid methanol systems | Low cell voltage and effi-ciency due to poor anode kinetics; Low power density; Lack of efficient catalysts for direct oxidation of metha-nol. Fuel and water crossover; Complex water manage-ment; High catalyst loading; High cost; Carbon dioxide (CO_2) removal system; Fuel toxic-ity | 35–60 | 3 | H |

| Direct ethanol | Solid Nafion*
 Alkaline media
 Alkaline-Acid media | Anode: Platinum–Ruthenium supported on carbon

 Cathode: Platinum supported on carbon | Graphite | Liquid ethanol–water solution | H^+ | Carbon monoxide (CO) | Ambient-120 | Compact size

 Environmentally-friendly fuel

 High fuel volumetric energy density

 Relatively low fuel toxicity

 Relatively higher gravimetric energy density

 •Easy fuel storage and delivery

 Simple thermal management | Low power density

 High sensitivity to carbon monoxide (CO)

 Low cell voltage and efficiency due to poor anode kinetics

 Lack of efficient catalysts for direct oxidation of ethanol

 High cost

 Fuel and water crossover | 20–40 | 2 | L |

Direct ethylene glycol	Solid Nafion* Anion exchange membrane (AEM)	Anode: Platinum supported on carbon / Cathode: Platinum supported on carbon	Graphite	Liquid ethylene glycol	H+	Carbon monoxide (CO)	Ambient-130	Advantages	Disadvantages	20–40	2	L
								Compact size	Low power density			
								High fuel volumetric energy density	Low cell voltage and efficiency due to poor anode kinetics			
								Low volatility due to low vapor pressure and high boiling point	Lack of efficient catalysts for direct oxidation of ethylene glycol			
								Easy fuel storage and delivery	Low fuel gravimetric energy density			
								Simple thermal management	Durability issues			
								Simple water management	High cost			
								Existence of distribution infrastructure	Fuel crossover			

| Microbial | Ion exchange membrane | Anode: Biocatalyst supported on carbon; Cathode: Platinum supported on carbon | N/A | Any organic matter (e.g., glucose, acetate, wastewater) | H^+ | Bacterial contamination of cathode | 20–60 | Fuel flexibility; Biocatalyst flexibility; No need for enzymatic catalysts isolation, extraction, and preparation; Relatively higher lifetime for biocatalysts; Capacity for self-regeneration of enzymes | Electron transfer mechanisms from the metabolism in the microorganisms to the fuel cell anode is problematic; Relatively lower energy density due to using some of the energy for the microorganisms activity; Very low power density; Low columbic yield; Inflexible operation conditions | 15–65[d] | 1 | M |

Enzymatic	Membrane-less Ion exchange membrane	Anode: Biocatalyst supported on carbon Cathode: Biocatalyst supported on carbon	N/A	Organic matters (e.g., glucose)	H$^+$	Foreign physical and/or chemical exposure to enzymatic catalyst	20–40	Capacity for miniaturization (e.g., for implantable medical micro-scale sensors and devices) Structural simplicity High response time	Rapid decay of enzymatic catalyst due to operation in foreign environmentHigh susceptibility to enzymatic poisoningElectron transfer mechanism from the reactive centers of the biocatalysts to the fuel cell electrodes is problematicImmobilizing the enzymes is problematicLow power densityVery low columbic yield-Low fuel flexibility-Inflexible operation conditions	30^d	1	M

| Direct carbon | Solid yttria-stabilized zirconia (YSZ) Molten carbonate Molten hydroxide | Anode: Graphite or carbon-based material Cathode: Strontium-doped lanthanum manganite (LSM) | N/A | Solid carbon (e.g., coal, coke, biomass) | O^{2-} | Ash Sulfur | 600–1000 | High electrical efficiency High volumetric energy density Fuel flexibility No PM, NOx, or SOx emissions Structural simplicity High capacity for carbon sequestration | Carbon dioxide (CO_2) emissions Rapid material corrosion and degradation Durability issues Sensitivity to fuel impurities Low power density | 70–90 | L | 2 |

Direct borohydride	Solid Nation* Anion exchange membrane (AEM)	Graphite	Anode: Gold, silver, nickel, or platinum supported on carbon Cathode: Platinum supported on carbon	Sodium borohydride (NaBH₄)	Na⁺	N/A	20–85	Compact size High fuel utilization efficiency High fuel gravimetric hydrogen content No carbon dioxide (CO₂) emissions Low toxicity and environmentally-friendly operation	Fuel crossover High cost-Low power density-Lack of analytical modeling techniques due to unknown borohydride oxidation reaction mechanismsExpensive catalyst-Chemical instability of membrane and catalystInefficient cathodic reduction reactionInefficient anodic oxidation reaction due to hydrogen evolution from hydrolysis of borohydride	40–50	M	2

Direct formic acid	Solid Nafion[a]	Anode: Palladium or platinum supported on carbon Cathode: platinum supported on carbon	N/A	Liquid formic acid (HCOOH)	H+	Carbon monoxide (CO)	30–60	Improved anodic oxidation reaction kinetics High fuel utilization efficiency Limited fuel crossover Easy fuel storage and delivery High power density No water required at the anode for oxidation reaction Compact size and structural simplicity	Fuel toxicity Components Corrosion issues Low fuel gravimetric and volumetric energy density High fuel cost Low temperature operation	30–50	L	1

[a]For the first typical electrolyte only.

[b]With 1 being lowest and 5 highest technological maturity relative to other fuel cell types.

[c]H: High; M: Moderate; L: Low.

[d]Columbic efficiency: The ratio of the coulombs transferred from the substrate to the anode to the coulombs produced if all the substrate is oxidized.

PAFCs are used for combined-heat-and-power (CHP) applications with high energy efficiencies. Molten carbonate fuel cells (MCFCs) and solid oxide fuel cells (SOFCs) are high-temperature fuel cells appropriate for cogeneration and combined cycle systems. MCFCs have the highest energy efficiency attainable from methane-to-electricity conversion in the size range of 250 kW to 20 MW, while SOFCs are best suited for base-load utility applications operating on coal-based gasses. Table 6 summarizes the main differences between the most common fuel cell types available in the market or still in the development stage.

CHARACTERISTICS AND FEATURES

Fuel cells have many inherent advantages over conventional combustion-based systems, making them one of the strongest candidates to be the energy conversion device of the future. They also have some inherent disadvantages that require further research and development to overcome them. We will elaborate on fuel cells' advantages and challenges in 3.1, 3.2, 3.3, 3.4, 3.5 and 3.6 and 3.7, 3.8, 3.9, 3.10, 3.11 and 3.12, respectively.

Reduced Harmful Emissions

The only products from a fuel cell stack fueled by hydrogen are water, heat, and DC electricity. And with the exception of controllable NOx emissions from high-temperature fuel cells, a hydrogen fuel cell stack is emissions-free. However, the clean nature of a fuel cell depends on the production path of its fuel (e.g. hydrogen). For instance, the products of a complete fuel cell system that includes a fuel reformation stage include greenhouse emissions (e.g., CO and CO_2). When the hydrogen supplied to the fuel cell is pure (i.e., not reformation-based hydrogen which is always contaminated with COx), the durability and reliability of the fuel cell significantly improve in comparison to when we run the fuel cell on reformation-based hydrogen. This is one of the most important advantages of fuel cells in comparison to heat engines, i.e., fuel cells are inherently clean energy converters that ideally run on pure hydrogen. This fact is actually pressingly driving researchers and the industry to develop efficient and renewable-based hydrogen generation technologies based on clean water electrolysis to replace the conventional reformation-based ones. Systems that integrate renewable-based hydrogen generation with fuel cells are genuinely clean energy generation and conversion systems that resemble what the energy industry is striving to achieve. It is worth mentioning that when we take into consideration the emissions from the fossil fuel reformation process, some heat engine systems appear to be less polluting than fuel cell systems (see Ref. [7]). For non-renewable energy based water electrolysis, the emissions and energy used for the electrolysis process make it more harmful to the environment than conventional combustion heat engines. Moreover, it is economically unfeasible since any fossil energy used for hydrogen production is going to be always more than the energy content of hydrogen. According to studies by Argonne National Laboratory [8], 3,000,000–3,500,000 BTUs of fossil energy are used for the production of 1,000,000 BTUs of hydrogen through fossil energy-based water electrolysis. This only stresses the significance of the aforementioned conclusions regarding using renewable-based water electrolysis for hydrogen production.

High Efficiency

The amount of heat that could be converted to useful work in a heat engine is limited by the ideal reversible Carnot efficiency, given by the following equation:

$$\eta_{Carnot} = \frac{T_i - T_e}{T_i}$$

(4)

where T_i is the absolute temperature at the engine inlet and T_e is the absolute temperature at the engine exit. However, a fuel cell is not limited by the Carnot efficiency since a fuel cell is an electrochemical device that undergoes isothermal oxidation instead of combustion oxidation. The maximum conversion efficiency of a fuel cell is bounded by the chemical energy content of the fuel and is found by (will be further discussed in Section 7.1):

$$\eta_{rev} = \frac{\Delta G_f}{\Delta H_f}$$

(5)

where ΔG_f is the change in Gibbs free energy of formation during the reactions and ΔH_f is the change in the enthalpy of formation (using lower heating value (LHV) or higher heating value (HHV)). Fig. 5 illustrates the efficiency of fuel cells and other energy conversion devices with respect to system power output [9]. In light vehicles, for instance, the efficiency of a fuel cell-powered car is nearly twice the efficiency of an internal combustion engine-powered car. Part of the reason why fuel cells have higher overall efficiencies compared to combustion-based energy conversion devices is illustrated in Fig. 6. The fact that the number of energy transformations that occur within a fuel cell stack is less than that of any combustion-based device, when the required output is electricity, plays a significant role. This is because losses are associated with each energy transformation process; thus, the overall efficiency of a system generally decreases as the number of energy transformations increases. As evident from Fig. 6, fuel cells have an advantage in the number of energy transformations over heat engines when the desired output is electrical power. However, fuel cells are on a par with batteries and heat engines in the number of energy transformations when the desired output is mechanical work.

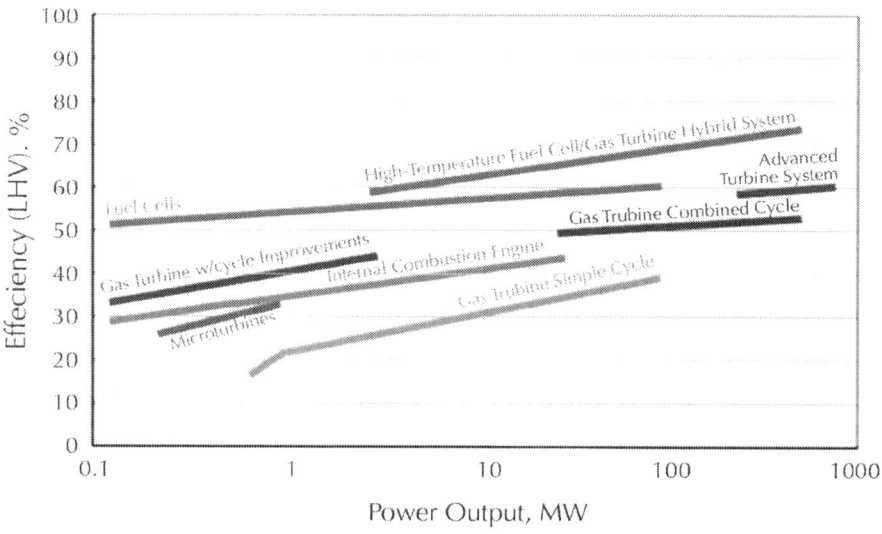

Figure 5: Efficiency comparison between fuel cells and other energy conversion devices with respect to system size [9].

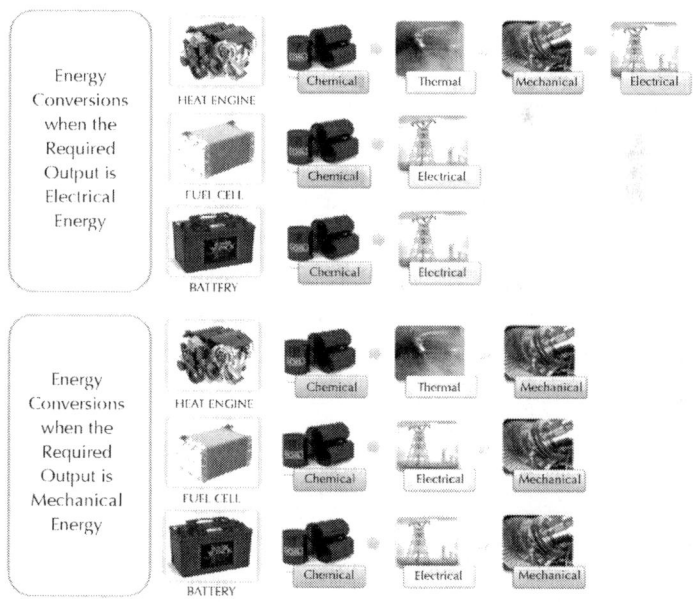

Figure 6: Energy transformations in fuel cells, batteries, and heat engines.

Modularity

Fuel cells have excellent modularity. In principle, changing the number of cells-per-stack and/or stacks-per-system (see Section 6) allows us to control the power output of any fuel cell system. Unlike combustion-based devices, a fuel cell's efficiency does not vary much with system size (see Fig. 5) or load factor. In fact, as opposed to conventional power plants, fuel cells have higher efficiencies at part loads compared to full loads (as will be shown in Section 7.5). This would prove advantageous in large-scale fuel cell systems that would normally run on part-load instead of full-load. Additionally, the high modularity of fuel cells means that smaller fuel cell systems have similar efficiencies to larger systems. This feature greatly facilitates the future integration of fuel cells (and hydrogen systems in general) in small-scale distributed generation systems, which hold a great potential in the power generation industry. It is worth noting; however, that reformation processors are not as modular as fuel cell stacks. This presents another reason to shift to renewable-based hydrogen production technology.

Prompt Load-following

Fuel cell systems generally have very good dynamic load-following characteristics [10] and [11]. This is partially due to the prompt nature of the electrochemical reactions that occur within a fuel cell. Again, when the fuel cell system includes a fuel reformation stage, the load-following ability of the system noticeably decreases as a result of the slower nature of the reformation process.

Static Nature

Due to its electrochemical nature, a fuel cell stack is a static silent device. This is a very important feature that promotes the use of fuel cells for auxiliary power and distributed generation applications in addition to portable applications that require silent-operation (as will be shown in Section 4). The fact that a fuel cell system has very few dynamic parts (and hence, almost no vibrations) makes fuel cells design, manufacturing, assembly, operation, and analysis simpler than that of heat engines. Nevertheless, for fuel cell systems that use compressors instead of blowers for the oxidant supply, noise levels can noticeably increase. As such, fuel cell designers tend to avoid using compressors due to their high parasitic load, noise production, cost, weight, volume, and complexity relative to fans and blowers. For instance, in a conventional urban bus, most of the noise is generated from the diesel engine. A fuel cell stack, on the other hand, is a completely silent device. As such, the noise level from a fuel cell bus could be significantly lower than a conventional bus provided that the fuel cell system's balance-of-plant (BoP) components are reasonably quiet. A study that compared a PAFC bus with diesel engine buses found the noise generated at 2.75 m from the former is at a maximum of 73 dB(A) while it ranges between 82 and 87 dB(A) for the latter [12]. The noise reduction in PAFC-based buses is greater than PEMFC-based buses [12]. This is due to the fact that, unlike PEMFCs, a PAFC operates at near-ambient pressures and does not require the use of a compressor. This, again, emphasizes the importance of developing efficient and simpler BoP components in a fuel cell system. A similar study found a PEMFC bus to generate 70.5 dB (A) compared to 76.5 dB(A) for a natural gas bus and 77.5 dB(A) for diesel buses at 10 m [13]. The static nature of a fuel cell also reflects on its low maintenance requirements in comparison to competing technologies such as heat engines, wind turbines, and concentrated solar power (CSP) plants.

Range of Applications and Fuel Flexibility

Fuel cells have diverse applications ranging from micro-fuel cells with less than 1 W power outputs to multi-MW prime power generation plants. This is attributed to their modularity, static nature, and variety of fuel cell types. This qualifies fuel cells to replace batteries used in consumer electronics and auxiliary vehicular power. These same properties also qualify a fuel cell to replace heat engines used in transportation and power generation. Fuel cells are also highly integrable to most renewable power generation technologies. Fuel cells that operate on low-temperature ranges require short warm-up times, which is important for portable and emergency power applications. While for fuel cells that operate on medium- to high-temperature ranges, utilization of waste heat both increases the overall efficiency of the system and provides an additional form of power output useful for domestic hot water (DHW) and space heating residential applications or CHP industrial-level applications. Fuels for a reformation-based fuel cell system include methanol, methane, and hydrocarbons such as natural gas and propane. These fuels are converted into hydrogen through a fuel reformation process. Alternatively, direct alcohol fuel cells (e.g. direct methanol fuel cells) can run directly on an alcohol. And even though fuel cells run best on hydrogen generated from water electrolysis, a fuel cell system with natural gas reformation also possesses favorable features to conventional technologies.

Fuel cells have been rapidly developing during the past 20 years due to the revived interest in them that started during the 1990s. However, they are still not at the widespread-commercialization stage due to many technical and socio-political factors, with cost and durability being the main hurdles that prevent fuel cells from becoming economically-competitive in the energy market. The main challenges are detailed as follows.

High Cost

Fuel cells are expensive. Experts estimate that the cost-per-kW generated using fuel cells has to drop by a factor of 10 for fuel cells to enter the energy market. Three main reasons behind the current high cost of fuel cell stacks are: the dependence on platinum-based catalysts, delicate membrane fabrication techniques, and the coating and plate material of bipolar plates. While from a system-level perspective, the BoP components such as fuel supply and storage subsystems, pumps, blowers, power and control electronics, and compressors constitute about half the cost of a typical complete fuel cell system. More specifically, whether renewable- or hydrocarbon-based, the current hydrogen production BoP equipment are far from being cost-effective. Technological advances in contaminate removal for hydrocarbon-based technologies are essential if the cost of fuel cell systems is to meet planned targets. Nevertheless, if fuel cells successfully enter the mass production stage, their costs are expected to significantly drop and become consumer-affordable due to the fact that manufacturing and assembly of fuel cells is generally less demanding than typical competing technologies, such as heat engines.

Low Durability

The durability of fuel cells needs to be increased by about five times the current rates (e.g., at least 60,000 h for the stationary distributed generation sector) in order for fuel cells to present a long-term reliable alternative to the current power generation technologies available in the market (more on this inSection 5). The degradation mechanisms and failure modes within the fuel cell components and the mitigation measures that could be taken to prevent failure need to be examined and tested. Contamination mechanisms in fuel cells due to air pollutants and fuel impurities need to be carefully addressed to resolve the fuel cell durability issue.

Hydrogen Infrastructure

One of the biggest challenges that face fuel cells commercialization is the fact that we are still producing 96% of the world's hydrogen from hydrocarbon reformation processes [14]. Producing hydrogen from fossil fuels (mainly natural gas) and then using it in fuel cells is economically disadvantageous since the cost-per-kWh delivered from hydrogen generated from a fossil fuel is higher than the cost-per-kWh if we were to directly use the fossil fuel. Thus, promoting renewable-based hydrogen is the only viable solution to help the shift from a fossil-based economy to a renewable-based, hydrogen-facilitated economy. Moreover, development of hydrogen storage mechanisms that provide high energy density per mass and volume whilst maintaining a reasonable cost is the second half of the hydrogen infrastructure dilemma. Any widely-adopted hydrogen storage technology will have to be completely safe since hydrogen is a very light and highly-flammable fuel that could easily leak from a regular container. Metal- and chemical-hydride storage technologies are proving to be safer and more efficient options than the traditional compressed gaseous and liquid hydrogen mechanisms. However, more research and development are needed to reduce the relatively high cost of the hydride storage technologies and to further improve their properties.

Water Balance

Water transport (fundamentally discussed in Section 7.3) within a fuel cell is a function of water entering with inlet streams, water generated by the cathodic reaction, water migration from one component to another, and water exiting with exit streams. Generally speaking, a successful water management strategy would keep the membrane well-hydrated without causing water accumulation and blockage in any part of the MEA or flow fields. As such, maintaining this delicate water balance inside a PEMFC over different operation conditions and load requirements is a major technical difficulty the scientific community is required to fully address [15]. Flooding of the membrane; water accumulation in the pores and channels of the GDL and flow fields; dryness of the membrane; freezing of residual water inside the fuel cell; dependence between thermal, gases, and water management; and humidity of the feeding gases are all subtle and interdependent facets in the water management of a PEMFC [15], [16] and [17]. Improper water management within a PEMFC leads to both performance loss and durability degradations [15] and [16] as a result of permanent membrane damage, low membrane ionic conductivity, non-homogeneous current density distribution, delamination of components, and reactants starvation. As such, water management strategies range from direct water injection to reactant gases recirculation [15]. The performance evaluation of a water management technique could be accomplished using empirical liquid water visualization or micro- and macro-scale numerical simulation [15]. Nonetheless, fundamental understanding and comprehensive models of water transport phenomena within a fuel cell are highly needed in order to develop optimized component designs, residual water removal methods, and MEA materials according to application requirements and operation conditions [16] and [17].

Parasitic Load

The parasitic load required to run the auxiliary BoP components reduces the overall efficiency of the system. This is clearly evident when the power required to run auxiliary components such as air compressors, coolant pumps, hydrogen circulation pumps, etc. is included in the efficiency calculations (as will be shown in Section 7.5). Additionally, the weight and size of fuel cell systems will need to be reduced in order for fuel cells to become compatible with on-board transportation applications and small-scale portable applications.

Codes, Standards, Safety, and Public Awareness

The lack of internationally-accepted codes and standards for hydrogen systems in general and fuel cells in particular has a negative reflection on the public's acceptance of hydrogen power solutions. Government officials, policy makers, business leaders, and decision makers would feel more reassured about supporting early-stage hydrogen power projects if general best-practices and consistent safety standards in the design, installation, operation, maintenance, and handling of hydrogen equipment were established. The general public needs to be convinced that hydrogen is similar to conventional fuels in certain aspects and different in other aspects. But overall, hydrogen does not pose a safety issue if properly handled and regulated, just like any other conventional fuel. Codes and standards for hydrogen systems could be made available by the continuous collection of more real-world data and initiation of more trial projects and lab experiments, a process that could be regulated by a professional society or a government initiative (in the US, the Safety, Codes, and Standards sub-program of the Department of Energy Hydrogen and Fuel Cells Program is attempting to take this vital role).

Table 7 summarizes the main advantages and disadvantages of fuel cells based on the previous discussions. As such, it become clear that fuel cells could play a major role in the portable sector where high energy and power densities and fuel flexibility are sought, in the stationary sector where reduced emissions and high modularity are sought, and in the transportation sector where high efficiencies and quick load-following are sought. These application areas and the fuel cell niche markets within them will be exhaustively covered in the following section.

Table 7: Summary of the main advantages and disadvantages of fuel cells

Advantages	Disadvantages
Less/no pollution	Immature hydrogen infrastructure
Higher thermodynamic efficiency	Sensitivity to contaminants
Higher part-load efficiency	Expensive platinum catalysts
Modularity and scalability	Delicate thermal and water management
Excellent load response	Dependence on hydrocarbons reformation
Fewer energy transformations	Complex and expensive BoP components
Quiet and static	Long-term durability and stability issues
Water and cogeneration applications	Hydrogen safety concerns
Fuel flexibility	High investment cost-per-kW
Wide range of applications	Relatively large system size and weight

APPLICATIONS AND MARKETS

Fuel cells hold promising potential to become competitive players in a number of markets due to their broad range of applications. And as a result of their high modularity, wide power range, and variation of properties among different types, fuel cells have applications ranging from scooters to large cogeneration power plants as fuel cells can theoretically be used for any energy-demanding application. Efforts towards the commercialization of fuel cells in the portable electronics, stationary power generation, and transportation sectors are well underway. In fact, worldwide shipments of fuel cells increased by 214% between the years 2008 and 2011 with fuel cells becoming an emerging competitor in the back-up power for telecommunication networks market [18], material handling market [19] and [20], and the airport ground support equipment market [21]. The global fuel cell industry market is expected to reach $19.2 billion by the year 2020 [22] with the United States, Japan, Germany, South Korea, and Canada acting as the flagship countries in the development and commercialization of fuel cells. Worldwide sales in the fuel cell industry are expected to grow 104% between 2010 and 2014 with annual installed power expected to exceed 1.5 GW by 2014 [9]. This was reflected on the employment growth in the fuel cell sector which grew at an average rate of 10.3% annually between the years 2003 and 2010 in the US [9]. Fig. 7, Fig. 8 and Fig. 9show the annual growth of fuel cells by application, region, and fuel cell type, respectively, between the years 2008 and 2012 with respect to both number of units shipped and MWs shipped [23].[1,2]

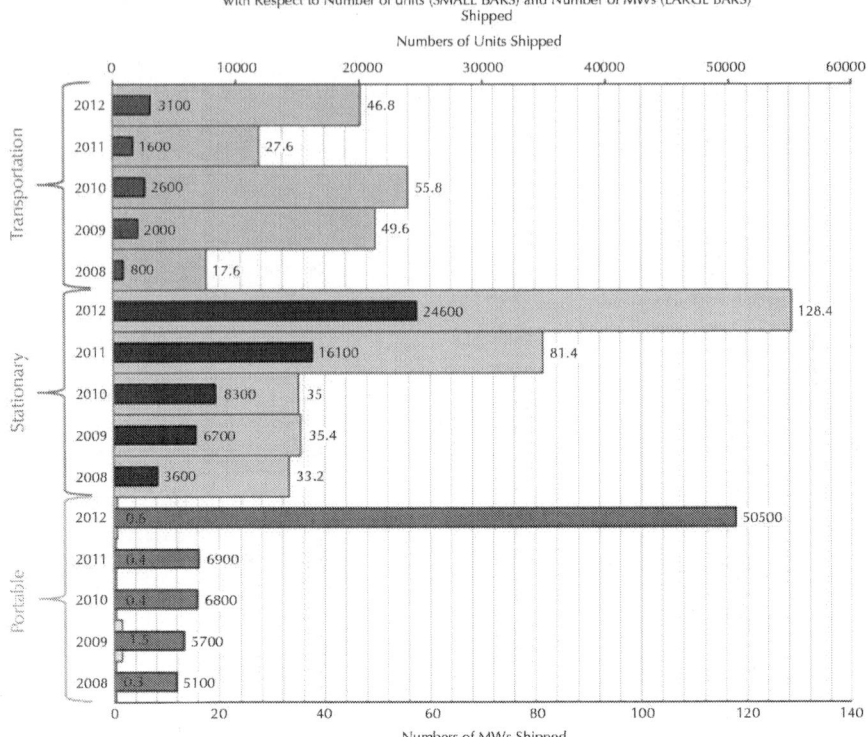

Figure 7: Annual growth of the fuel cell industry between 2008 and 2012 by application. Source: Data Sourced from [23].

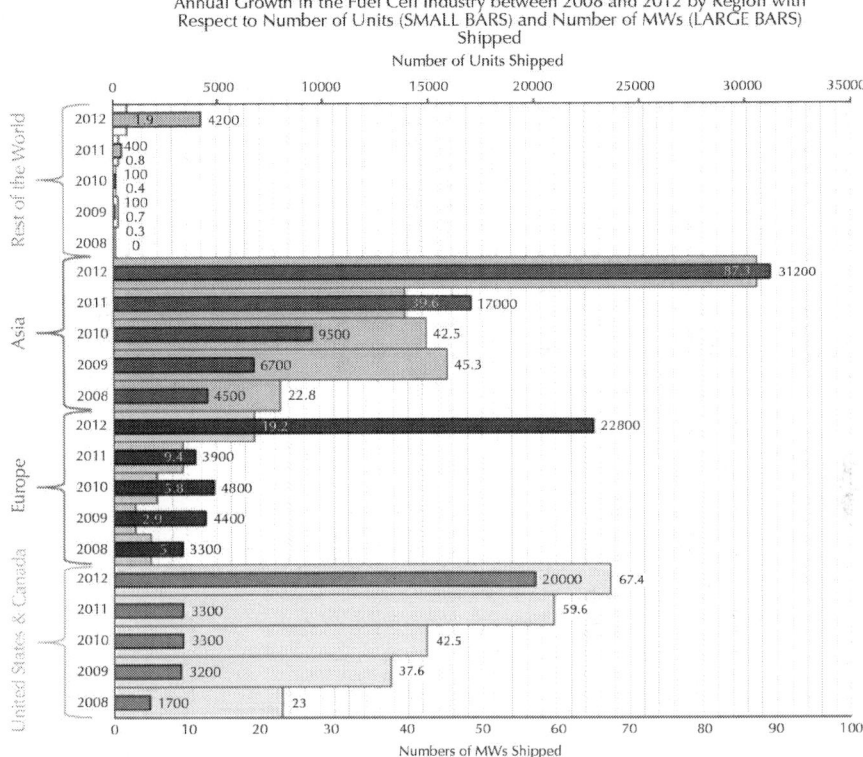

Figure 8: Annual growth of the fuel cell industry between 2008 and 2012 by region. Source: Data Sourced from [23].

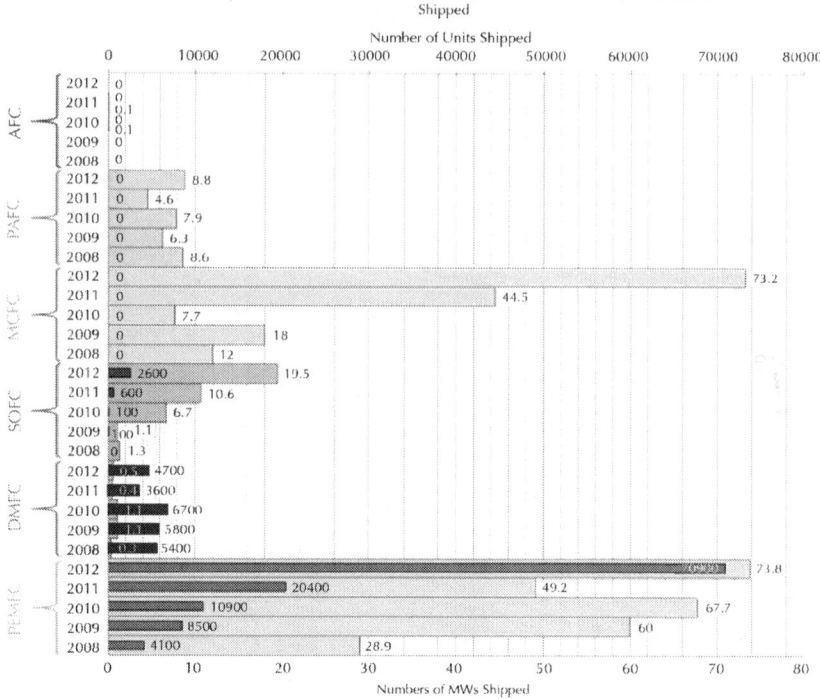

Figure 9: Annual growth of the fuel cell industry between 2008 and 2012 by fuel cell type. Source: Data Sourced from [23].

Portable Applications

Portable applications for fuel cells are mainly focused on two main markets. The first is the market of portable power generators designed for light outdoor personal uses such as camping and climbing, light commercial applications such as portable signage and surveillance, and power required for emergency relief efforts. The second is the market of consumer electronic devices such as laptops, cell phones, radios, camcorders, and basically any electronic device that traditionally runs on a battery. Portable fuel cells typically have power ranges between 5 and 500 W, with micro-fuel cells having power outputs less than 5 W and more demanding portable electronics reaching the kW-level. Unlike stationary fuel cells, portable fuel cells could be carried by an individual and used for a variety of applications. The modularity and high energy density of fuel cells (5–10 times higher energy density than a typical rechargeable battery) make them strong potential candidates for future portable personal electronics. Moreover, portable military equipment is another growing application for portable direct methanol fuel cells (DMFCs), reformed methanol fuel cells (RMFCs), and PEMFCs due to their silent operation, high power and energy density, and low weight compared to current battery-based portable equipment [24] and [25]. In addition to lower weight and higher energy density, the fact that fuel cells do not require recharging from an electricity source makes them more favorable in comparison to batteries in the future portables market. However, their cost and durability are yet to meet set targets. Other rapidly-growing markets in the portable sector include portable battery chargers in addition to miniature demonstration and educational remote control (RC) vehicles, toys, kits, and gadgets by manufacturers such as Horizon [26] and Heliocentris [27]. The vast number of portable equipment being integrated with fuel cells has made roughly half the total number of fuel cell units shipped in 2008 fall into the portable sector category [28], even though on a MW-level, the portable sector accounts for less than 1% of the worldwide fuel cell shipments [29] in the years 2008–2011. However, issues of heat dissipation, emissions dissipation, noise, integrated fuel storage and delivery, shock and vibrations endurance, response time to sharp and repeated demand fluctuations, operation under various operation conditions, tolerance to air impurities, reusability and recyclability of fuel containers, and area exposed to oxygenated air need to be addressed before any serious advances in the portable sector could be made.

Stationary Applications

Fuel cells can play an integral part in the residential, commercial, and industrial stationary power generation sectors. They are utilized for both grid-independent (also known as stand-alone) and gird-assisted power supply. Stationary fuel cell applications include emergency back-up power supply (EPS) (also known as uninterrupted power supply (UPS)), remote-area power supply (RAPS), and distributed power or CHP generation. The stationary fuel cells market accounts for about 70% of the annual fuel cell shipments on a MW-level [29].

Emergency Back-up Power Supply (EPS)

Due to their high energy and power densities, high modularity, longer operation times (2–10 times longer than currently-used lead-acid batteries), compact size, and ability to operate under harsh ambient conditions, fuel cells are becoming an encouraging alternative for batteries in the EPS market, especially in the telecommunications market, with PEMFCs and DMFCs as the dominantly-chosen fuel cell types [18]. Due to the fact that the EPS market requires high reliability but not necessarily high operational lifetimes, fuel cells found EPS to be one of its most successful markets. Other fuel cell EPS markets include hospitals, data centers, banks, and government agencies. In all these markets, the continuation of a power supply (typically between 2 and 8 kW) is critical, when grid power is unavailable.

Remote-area Power Supply (RAPS)

In gird-isolated locations, such as islands, deserts, forests, remote technical installations, holiday retreats, and remote research facilities, providing power could be problematic. Such locations fall under the remote-area power supply (RAPS) category. Usually, providing power to rural and urban off-grid locations using RAPS solutions is more economical than extending electric grid power lines. This is especially true for rural areas where the geographical nature of rough terrains (forests, mountains, etc.) makes grid extensions an unrealistic approach. In fact, experts state that extending electric grid lines to rural locations is more expensive when compared to urban locations due to the low load densities in rural areas, high transmission losses for distant areas, and the high cost of the required rural infrastructure that needs to accompany the grid extensions [30]. Off-grid rural areas, especially in developing countries, are a typical example that could significantly benefit from RAPS power generation solutions [31], [32], [33], [34] and [35]. Off-grid households in urban areas, as well, use various RAPS solutions for their energy needs [36] and [37]. Currently, RAPS energy solutions are hydrocarbon-based [37], renewable-based [31], [34], [35] and [38], or a combination of both [36]. CHP expansion for current RAPS solutions is another option being investigated in order to increase overall system efficiency and add another useful power output, especially for residential urban RAPS locations where the additional increase in capital cost could be tolerated [39]. Similarly, light off-grid industrial and commercial applications, such as technical installations, water pumps, and medical centers use the same RAPS solutions [31], [34] and [38]. Classically, diesel engines have been the energy conversion devices of choice for RAPS. However, diesel engines have high carbon footprints and noisy operation. Fuel cell systems that contain a natural gas or light hydrocarbon reformation component could serve as an alternative for RAPS [37]. However, delivering natural gas or the hydrocarbon through pipelines or other means makes this alternative less appealing for rural and remote locations. Hybrid and integrated energy systems that couple a renewable energy source (such as hydro, biomass, solar, wind, etc.), depending on available natural resources, with a storage mechanism (lead-acid batteries, lithium-ion batteries, hydrogen systems, etc.) provide a more sustainable and autonomous solution for RAPS [31], with solar-PV being the preferred renewable RAPS power generation solution. China and India are leading the efforts in the utilization of integrated renewable-based systems for the RAPS of rural areas [30]. In fact, by 2030, it is expected that PV-based RAPS systems will amount to 130 GW, with about half of the installed capacity to be utilized in industrial and commercial applications and the other half in the residential sector [40]. A hydrogen system coupled with a renewable energy source is typically composed of a water electrolyzer, hydrogen storage mechanism, and a fuel cell. The utilized hydrogen system would compensate for the intermittency problem most renewable sources suffer from, making the system in its totality both reliable and sustainable. However, PV systems in RAPS applications are typically coupled with batteries for energy storage. Thus, once again, fuel cells are competing with batteries for the same market share. In a comparison conducted by Perrin and Lemaire-Potteau [40] between different types of batteries and hydrogen systems for RAPS applications, it was found that batteries are superior to a hydrogen system with respect to their thermoeconomic performance due to the high cost and low overall efficiency of a hydrogen system. Nevertheless, a RAPS system that uses a hydrogen system for seasonal storage could prove to be favorable. Zoulias and Lymberopoulos [41] found that the complete replacement of batteries with a hydrogen system in a real PV-batteries RAPS system is technically possible, but not yet economically feasible due to the high capital cost for the hydrogen system equipment. Munuswamy et al. [34] concurred the findings of Zoulias and Lymberopoulos [41], stressing that cost reductions in hydrogen systems components are vital for the penetration of renewable-based, hydrogen-facilitated technologies in RAPS. Cost reductions of about 50% for the water electrolyzer, 40% for the hydrogen storage tanks, and a cost of 300/kW for the fuel cell system would make the hydrogen system economically superior to conventional batteries [41]. Khan and Iqbal [36] reached a similar conclusion for a wind-based system that uses a hydrogen system for energy storage. They concluded that in order for the wind–hydrogen system to be advantageous in comparison to the optimal diesel—wind-battery option in the region they conducted their study, the fuel cell cost will have to be reduced by 85% of the current value. With nearly two billion people living without grid electricity, the development of sustainable and reliable RAPS energy solutions becomes a necessity in which fuel cells and hydrogen systems could make a substantial difference if their cost, overall efficiency, and durability issues are resolved.

Distributed Power/CHP Generation

Fuel cells could serve as the means to make the shift from large centralized power generation to decentralized distributed generation. Due to their static nature, lower emissions, excellent load-following, and high efficiency; fuel cells could be used for residential electric power or CHP distributed generation either on a household basis [42], [43] and [44] or a larger residential blocks basis [45] and [46]. In fact, it is estimated that by 2020, fuel cells could penetrate 50% of the world distributed generation market if cost and durability targets are met [32]. This is a market in which Japan is the current leader where thousands of households are already relying on distributed CHP fuel cell systems for their power and heating needs [29]. A residential CHP fuel cell system could range from a few kilowatts to a few megawatts depending on the targeted basis load. A residential CHP fuel cell system will be able to provide electric power, space heating, and domestic water heating requirements. Cooling could also be added to power generation and heating (known as combined cooling, heating, and power (CCHP) systems) if an absorption chiller, thermally-driven heat pump, or an appropriate technique (see Ref. [47]) is integrated with the system to utilize the waste heat of the fuel cell stack in a dual-mode heating/cooling cycle [44], [46], [48] and [49]. CHP and CCHP systems could reach overall efficiencies as high as 80% [42] and [49]; however, further studies to resolve the technical challenges and reduce capital cost in addition to funding for experimental validation remain highly needed. PEMFCs and PAFCs are the best candidates for household basis CHP generation, while high-temperature fuel cells (HTFCs) are more suitable for the larger residential block basis CHP generation. Fuel cell systems used for CHP generation could be designed to either be grid-independent or grid-assisted; however, system complexity and cost increase for grid-independent since the system will have to meet dynamic load fluctuations, which will be more felt for a household basis unit. Oversized fuel cell systems and the integration with battery banks/ultracapacitors are two solutions for this issue; however, both come at a higher cost and system complexity. On the other hand, gird-assisted systems export electricity to the grid during low load demands and import electricity from the grid during peak load demands. For both cases, a thermal storage mechanism is a requirement for an effective CHP system. As for harmful emissions, on a life-cycle assessment basis, MCFCs and PAFCs running on natural gas in CHP cycles emit 78–88% less NOx emissions, about 60% less particulate matter-10 emissions, and 90–99% less CO emissions in comparison to other combustion-based distributed generation CHP technologies [50]. Aside from residential CHP, combined fuel cell cycles for both energy and chemicals generation in industrial processes [51] and [52] are being developed. It is worth noting that, regardless of the application for stationary CHP fuel cells, the most challenging technical difficulty remains to increase the lifetime of the fuel cell system to achieve a target of 80,000 h.

Transportation Applications

The transportation industry is one of the main powerhouses in the development of clean energy technologies. This is due to the fact that the transportation industry is responsible for 17% of the global greenhouse gas emissions every year [53]. The industry's outlook is to invest in technologies that would offer both significant reductions in harmful emissions and better energy conversion efficiencies. Accordingly, the current complete dependence on combustion-based technologies that utilize fossil fuels in heat engines makes the development of environmentally-benign transportation alternatives a necessity rather than an option. This is where fuel cells come into the picture. Fuel cells offer the transportation industry near-zero harmful emissions without having to compromise the efficiency of the vehicle's propulsion system. In fact, fuel cells have demonstrated efficiencies (from 53% to 59%) that are almost twice the efficiencies of conventional internal combustion engines [54]. When we add advantages such as static operation, fuel flexibility, modularity, and low maintenance requirements; fuel cells become an ideal future alternative for current combustion engines. That is, if durability, cost, hydrogen infrastructure, and technical targets are met on-schedule. This is why using fuel cells in various means of transportation, with a focus on light-duty passenger cars, has been one of the main drivers for fuel cell R&D in the past decade. Japan, for example, has announced an aggressive development plan to deploy two million fuel cell electric vehicles (FCEVs) with one thousand hydrogen refueling stations by 2025 [9]. The share of transportation-related fuel cell shipments worldwide constituted about 35% and 25% of the total fuel cell systems shipped on a number of units basis and a MW-level basis, respectively, in 2010 [29], with PEM as the main fuel cell type chosen. We will classify fuel cell applications that fall under the transportation category into the following markets: auxiliary power units (APUs), light traction vehicles (LTVs), light-duty fuel cell electric vehicles (L-FCEVs), heavy-duty fuel cell electric vehicles (H-FCEVs), aerial propulsion, and marine propulsion. Most of the efforts in the transportation area are focused on APUs and L-FCEVs, as will be observed in the following discussions.

Auxiliary Power Units (APUs)

An on-board auxiliary power unit (APU) is used for the generation of non-propulsive power in any vehicle. Unlike portable power generators that could be used on-board of recreational vehicles (RVs), boats, etc., an APU is built into the vehicle. The load an APU has to meet could range from less than a kilowatt to loads up to 500 kW on large commercial airplanes. This is why separation of the main propulsion system from the APU is a good strategy to optimize the overall vehicle energy consumption. An APU provides power for air conditioning, refrigerating, entertainment, heating, lighting, communication, and any electrical appliances in any car, boat, ship, locomotive, airplane, truck, bus, submarine, spaceship, military vehicle, or any other vehicle with on-board energy needs. However, leisure yachts, planes, and cars [55]; heavy-duty trucks[56] and [57]; utility and service vehicles; law enforcement vehicles; and refrigeration vehicles present the most promising markets for APUs due to their high on-board electrical energy demand [58], [59] and [60]. Leisure and recreational vehicles in particular have extensive on-board APU uses that could mount to more than 38 kW, as seen from Table 8 that lists the APU demands for a typical luxury passenger vehicle in 2000[61]. With the increased number of on-board comfort features and electrical appliances in the vehicle industry, the demand an APU has to meet is always growing. Fig. 10 shows the structure of a 450 W liquefied petroleum gas-based (LPG-based) PEM fuel cell system with an on-board reformation unit used on a leisure yacht as an APU [55]. Three main APU strategies are currently dominating the market; on-board batteries, on-board hydrocarbon generators, and drawing power from the main propulsion system. Fuel cells as APUs produce significantly less emissions, cause no acoustic pollution, have short start-up times, and have higher efficiencies. To get a feeling of the possible reduction of emissions if fuel cells are used as APUs, a study by Gaines and Hartman [62] concluded that using on-board fuel cell APUs instead of the two most commonly used APUs in heavy-duty trucks in the US – drawing power from the main diesel propulsion engine and separate on-board diesel generators – could decrease particulate matter-10 emissions by up to 65%, NOx emissions by up to 95%, and CO_2 emissions by more than 60%. Other studies show that truck idling (the practice of keeping the truck's engine on when the truck is not moving for HVAC purposes, electric appliances operation, or keeping the engine ready) accounts for 20–40% of the overall engine running time (about 6 h a day) in heavy-duty trucks that draw power from the engine for APU, during which the energy efficiency is only 3% [59]. Accordingly, a truck on idle mode would consume roughly one gallon of diesel fuel per hour. This represents a great waste of energy and fuel, an extra burden on the engine, a major source of harmful emissions (which led to certain legislations that limit/ prohibit truck idling in populated areas), and an impractical engineering practice. Currently, PEMFCs, DMFCs, and SOFCs are the fuel cell types being developed for APU applications with pure hydrogen, natural gas, LPG, gasoline, methanol, and diesel as potential fuels. Other more energy-intensive vehicles, such as commercial airplanes and cargo ships, require APUs with high energy ratings, for which high-temperature fuel cells (SOFCs and MCFCs) become better candidates.

Table 8: APU demands in a luxury passenger vehicle [61]

Accessory	Power demand (W)
Rear wiper	90
Infotronics	100
Windshield pump	100
Heated steering wheel	120
Power sunroof	200
Truck closer	200
Windshield wipers	300
Air pump	400
Power door locks	400
Engine coolant pump	500
Antilock braking system (ABS) pump	600
Lights	600
Power windows	700
Electric fan	800
Rear defrost	1000
Power seats	1600
Steer by wire	1800
Brake by wire	2000
Heated front seats	2000
Heated windshield	2500
Catalyst heating	3000
Electromechanical valve control	3200
Air conditioning	4000
Active suspension	12,000

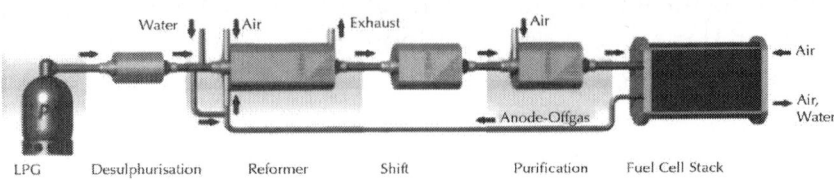

Figure 10: 450 W PEM fuel cell systems with on-board reformation for APU on a Leisure Yacht [55].

Light Traction Vehicles (LTVs)

Light traction vehicles (LTVs) include scooters, personal wheelchairs, electric-assisted bicycles, airport tugs, motorbikes, golf carts, etc. in addition to material handling vehicles and equipment. Material handling vehicles and equipment include forklifts, tow trucks, pallet trucks, etc. and falls under the LTVs category. Forklifts have been the most successful demonstration of fuel cells in the transportation sector, and one of the most successful demonstrations for fuel cells overall. Forklifts and other material handling vehicles and equipment are exhaustively used in the warehousing and distribution industry, with nearly 2.5 million forklifts in-operation in North America [63]. Most forklifts use rechargeable lead-acid batteries (usually with regenerative braking energy recovery) or combustion engines (either compression ignition engines running on diesel or spark ignition engines running on gasoline, LPG, compressed natural gas, or propane). However, since fuel cells require only 2–5 min for refueling from a fueling station in contrast to recharging or changing batteries which takes 15–30 min (which increases operational efficiency), have longer operation cycles in contrast to battery cycles which usually last less than 8 h (again, increasing operational efficiency), are less sensitive to the ambient temperature (especially in refrigerated warehouses) in contrast to batteries, do not self-degrade with charge and discharge cycles in contrast to batteries, require much less space for refueling stations in contrast to battery charging and changing space, have significantly less harmful emissions compared to combustion-based engines, can operate indoor or outdoor in contrast to many combustion-based forklifts that cannot run indoors, have high efficiencies, have excellent load-following dynamics, and require low maintenance; they have a huge potential to gradually replace conventional forklifts. It is important to notice that liquid hydrogen/fuel feeding systems that use a fueling station are much more practical than on-board hydrogen reformation or generation systems that tend to be bulky and could significantly slow down the response time. Fig. 11 shows the fuel cycle greenhouse gas (GHG) emissions of forklifts that use diesel engines, gasoline engines, LPG engines, batteries with charging from the average US electric grid mix, batteries with charging from the average California electric grid mix, batteries with charging from a simple natural gas steam cycle, batteries with charging from a combined natural gas cycle, fuel cells with hydrogen from steam reforming of methane (the main component in natural gas), fuel cells with hydrogen from coke oven gas (COG), and fuel cells with hydrogen from wind energy. The fuel cycle GHG emissions includes in the calculations the upstream emissions that accompanied converting primary energy sources into the usable forms in the forklifts, the point of use emissions (which is applicable only to combustion engine-based forklifts), and emissions from hydrogen compression for fuel cells [8]. Around 1300 fuel cell-powered forklifts are operative in the US market today [64]. Fuel cell forklifts are typically operated on 5–20 kW PEMFCs, with few models running on DMFCs, coupled with ultracapacitors for instantaneous power response support. Plug Power remains as the biggest player in the fuel cell forklift market since the company is solely focused on the material handling market. Fuel cell-powered scooters [65] and [66] and fuel cell-powered electric-assisted bicycles [67] are also being developed and expected to have a good share of the future fuel cell market since they help avoid traffic congestions, are ideal for short- and medium-distances trips, are environmentally-benign, and do not consume expensive hydrocarbon fuels. It is important to note that motorcycles, scooters, and electric-assisted bicycles are differentiated in descending order based on the power requirement, total weight, available speed, and travel distance. Scooters have power requirements between 4 and

6 kW with travel distances up to 200 km, while electric-assisted bicycles require less than 1 kW of power with a travel distance less than 1 km. Electric-assisted bicycles use a combination of human pedaling and electric motors that run on batteries. For the same reasons listed in our discussion on forklifts (emissions, charging time, operation duration, etc.), fuel cell-powered scooters and electric-assisted bicycles are favorable to combustion- and battery-based ones. Other fuel cell-powered LTVs such as light carts [68] and personal wheelchairs [69] have also been demonstrated.

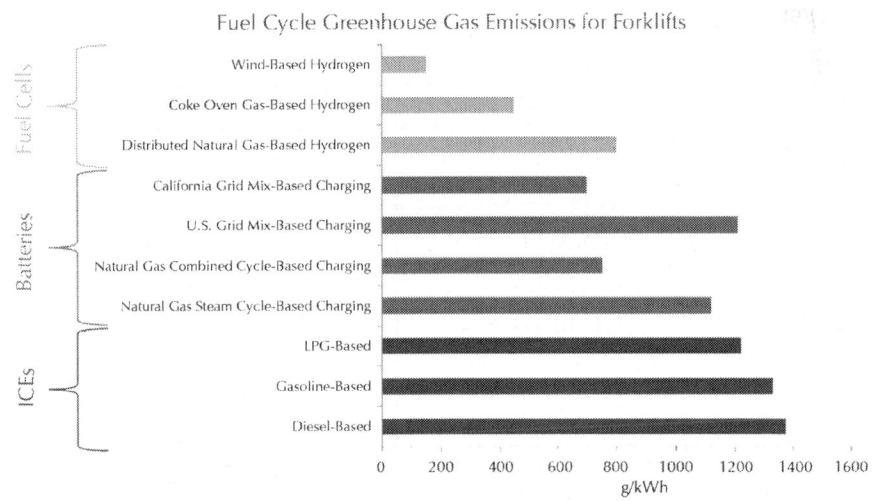

Figure 11: Fuel cycle GHG emissions for ICE-, battery-, and fuel cell-powered forklifts. Source: Data Estimated from [9].

L-FCEVs

Light-duty fuel cell electric vehicles (L-FCEVs) utilize a fuel cell for the propulsion system. When compared to the current internal combustion engine-based vehicles, L-FCEVs provide quitter operation (due to the static nature of fuel cells), more efficient energy use (nearly twice the efficiency of internal combustion systems, as mentioned earlier, and the potential for about 30%

higher well-to-wheel efficiency [70]), significantly less point-of-use GHG emissions (and the potential for near-zero lifecycle GHG emissions if renewables are used for the production of hydrogen), and more vehicle design and packaging flexibility. When compared to light-duty battery electric vehicles (L-BEVs), L-FCEVs provide a longer range, shorter refueling time (less than 2 min), better tolerance of cold weather, and lighter weight. However, lifecycle cost and stack durability limitations are the reasons L-FCEVs have not become commercially-available yet. Other technical barriers that need to be resolved are related to total system weight and size; air compression systems; start-up in very cold weather and under frozen conditions; heat dissipation; catalyst tolerance to voltage cycling; stack endurance of frequent start-stop cycles; bipolar plates weight; on-board hydrogen storage; membrane humidification; and hydrogen safety standards. Due to the inherent advantages (related to dynamics response, operation temperature, system size, etc.) PEMFCs hold over other fuel cell types, hydrogen PEMFCs are the most used fuel cells in L-FCEVs research, development, and demonstration efforts. General Motors, Toyota, Mazda, Daimler AG, Volvo, Volkswagen, Honda, Hyundai, Nissan, and other major car manufacturers are steadily progressing for the near-term commercialization of L-FCEVs that utilize a fuel cell for the main propulsion system through an electric motor connected to the fuel cell system. February 26, 2013 witnessed a major milestone in the history of the fuel cell industry as Hyundai announced the completion of its first assembly line mass-produced L-FCEV, the ix35 [71]. The Korean automobile manufacturer will deliver its first 17 L-FCEVs to customers in Denmark and Sweden in a few months. The company aims to have 1000 of its ix35's running on the streets of Europe by 2015 with a cost of $50,000 per car. The vehicle has a range of nearly 600 km and a top speed of 160 kmph using its hydrogen/air fuel cell system, which could run at temperatures as low as −25 °C, coupled with a Li-ion battery for regenerative braking energy recovery. The main typical components of a L-FCEV include fuel cell stacks; cooling systems for the stack, motor, and transmission; high-pressure hydrogen storage tanks (or any compact, light, and energy-dense storage mechanism); electric motor; main power control unit; high-voltage batteries and/or ultracapacitors for regenerative braking energy recovery (which could increase the driving range by 5–20% [70]) and response to transient power surges; air and hydrogen supply systems; power conditioning electronics; and other auxiliary BoP components.Fig. 12 shows the conceptual design of a typical future L-FCEV [72].

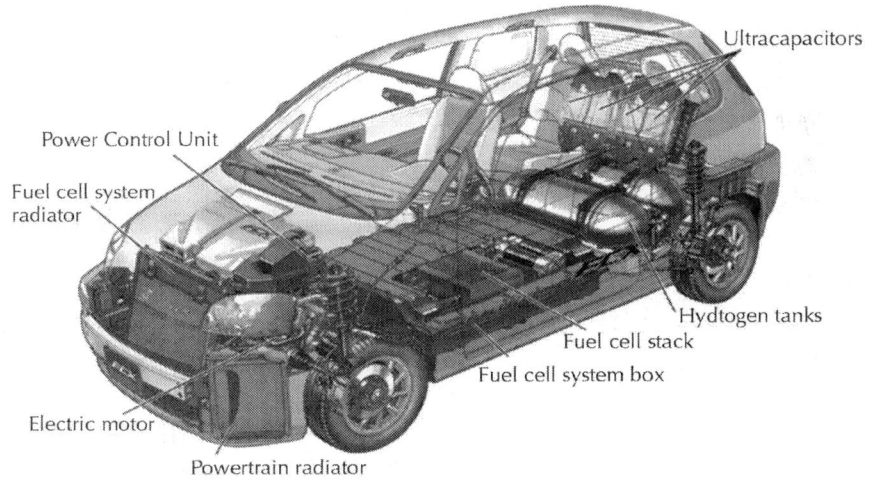

Figure 12: Conceptual design of a future L-FCEV based on the Honda 2005 FCX model [72].

The main competition for L-FCEVs comes, as one would expect, from L-BEVs, with nickel metal-hydride (NiMH) batteries currently dominating the L-BEVs market and lithium-ion (Li-ion) batteries steadily progressing to be the future choice in L-BEVs [73]. Both L-FCEVs and L-BEVs have advantages and disadvantages that are highly dependent on the primary energy source (fossil, renewable, biomass, etc.), energy conversion chain (e.g., hydrogen production, transport, and storage mechanisms), and design requirements (maximum speed, driving range, passenger capacity, etc.). The results of a study conducted by Campanari et al. [74] to compare the energy and environmental characteristics of L-BEVs and L-FCEVs imply that using light-duty hybrid electric vehicles (L-HEVs) that utilize both batteries and fuel cells in an attempt to combine their advantages might be an interim solution until one of the technologies (both PEMFCs and Li-ion batteries are still immature and going through rapid technological advances) shows obvious superiority over the other. For example, the power demand could be distributed between the fuel cell and the battery where one meets the bulk average power demand while the other meets the transient acceleration fluctuations, which would improve the fuel cell's durability due to avoidance of repeated voltage cycling and help avoid the design of an oversized fuel cell system. A comprehensive study that compared state-of-the-art advanced Li-ion, Ni-MH, and deep discharge lead-acid batteries with PEM fuel cells using computer simulations by Thomas [75] concluded that L-FCEVs are superior to L-BEVs for any driving range greater than 160 km with respect to vehicle's mass, storage volume, incremental vehicle cost, incremental lifecycle costs, GHG emissions, refueling time, and well-to-wheel energy efficiency when natural gas or biomass are the primary energy source. Fig. 13 presents modeling estimates for the amount of greenhouse gas emissions and the level of petroleum use for future light-duty mid-size passenger cars that utilize hydrogen fuel cells, batteries, diesel fuel, gasoline fuel, cellulosic ethanol, corn ethanol, and natural gas for the propulsion system [76]. For each type there are some technological variations, as seen from the figure. For example, the source of the electricity used to charge the batteries (grid electricity or renewable-based electricity) for a future mid-size L-BEV makes a noticeable impact on the GHG emissions and level of petroleum use. The modeling estimates have been carried using a projected technological state to 2035–2045 that does not include life-cycle effects of vehicle manufacturing and infrastructure (for a full record of the assumptions and references used for the models and estimates refer to [76]). In the figure, the renewables referred to are carbon-free technologies such as solar, wind, and ocean energy. Cars that use batteries, hydrogen fuel cells, cellulosic ethanol, and corn ethanol have much lower GHG emissions when compared to other technologies. While for the petroleum use, batteries, hydrogen fuel cells, and natural gas have the lowest levels. All renewable-based technologies show superiority with respect to both GHG emissions and petroleum use when compared to other technological variations within each type. However, renewable-based L-FCEVs remain dependent on simultaneous technological advances in hydrogen production, storage, and delivery; making natural gas-based and biomass-based L-FCEVs a more forthcoming and practical light-duty transportation alternative for the short-term future.

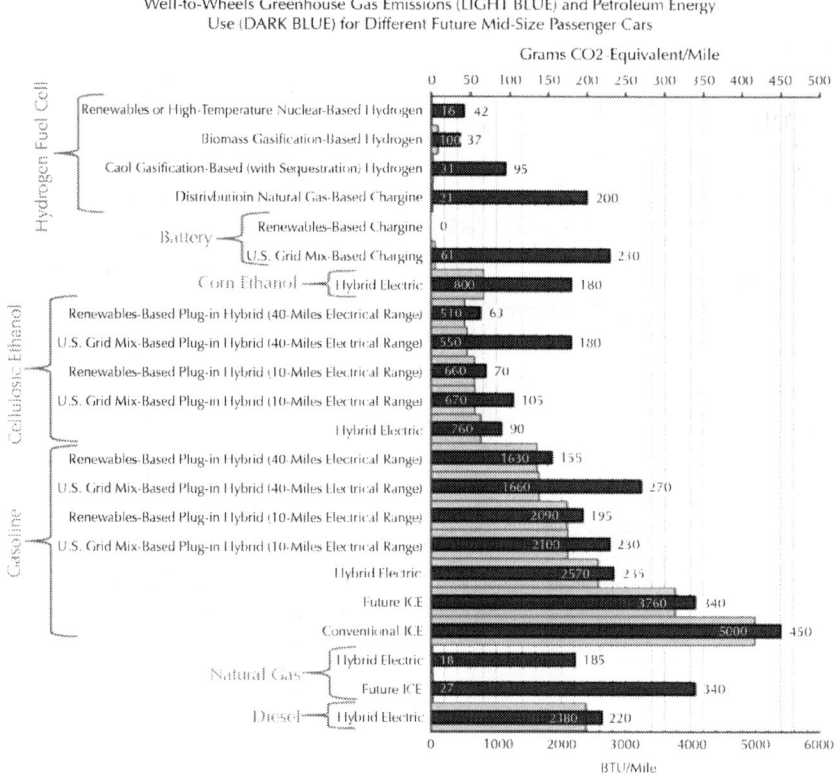

Figure 13: Well-to-wheel greenhouse gas emissions and petroleum energy use estimates for future mid-size passenger cars using different propulsion technologies. Source: Data Sourced from [76].

The hydrogen fuel required to run a L-FCEV is usually generated off-board and distributed in dedicated fueling stations where hydrogen is then dispensed to the hydrogen storage system on-board of the vehicle. On-board hydrogen storage is one of the biggest challenges and most active research areas for the commercialization of FCEVs. Using compressed hydrogen, liquid hydrogen, metal hydrides, chemical hydrides, and other novel storage technologies are all being evaluated. Unlike for stationary applications, the constraints for the hydrogen storage system in transportation applications are more demanding and inflexible. In addition to the low cost, high efficiency, and low parasitic load (e.g. for compression, cooling, or discharging) constraints; gravimetric energy density (the leading issue with metal-hydrides), volumetric energy density (the leading issue with compressed gas), collision safety requirements, fitting into the vehicle's space and shape, and system complexity are additional constraints for a hydrogen storage system on-board of most means of transportation. In fact, a recent study by Ahluwalia et al. [77] shows that none of the currently-available hydrogen storage mechanisms meet the long-term targeted characteristics for a future L-FCEV. However, the study concludes that cryo-compressed hydrogen storage, ammonia borane chemical storage, and alane metal hydrides hold the most potential for meeting future storage targets; given that certain technical issues are resolved. It is worth noting that using on-board reformation (or any other hydrogen production method) for hydrogen generation is still impractical due to size, weight, start-up time, and safety limitations. Nevertheless, with advances in reformation and hydrogen production technologies, on-board generation could become feasible.

H-FCEVs

Heavy-duty fuel cell electric vehicles (H-FCEVs) include buses, heavy-duty trucks, locomotives, vans, utility trucks, service fleets, etc. that utilize a fuel cell for the electric propulsion system. With more than 30 fuel cell buses currently deployed in Western Europe [78] and another 25 in the United States [79] in 2012; fuel cell electric buses (FCEBs) are becoming one of the best public demonstration tools and R&D data sources in the fuel cell transportation industry. And with public transportation becoming more desirable for urban communities in order to help reduce harmful emissions and avoid traffic congestions from the use of personal cars; buses are one of the most appealing applications for fuel cell technology in order to progress towards clean public transportation. FCEBs provide very low emissions (zero point-of-use emissions for buses that only use hydrogen fuel cells and batteries) when compared to conventional diesel-based buses, silent operation, and higher efficiency compared to other combustion-based busses, making them favorable to the public and the policy makers. While when compared to other fuel cell-powered vehicles, buses provide more flexible design and packaging, less complex hydrogen infrastructure requirements since bus routes are usually fixed, and more flexible weight and size constraints for the hydrogen storage system. This is why numerous programs funded by governments and the private sector were found with the aim of deploying FCEBs in the US, Canada, Western Europe, Japan, China, Australia, and South America. Key examples include the Sustainable Transport Energy Programme (STEP) initiative in Australia [80], the Clean Urban Transport for Europe (CUTE) and HyFLEET:CUTE (which is the world's largest FCEBs demonstration program with 33 FCEBs deployed) projects in Europe [81], the Hydrogen Fuel Cell Demonstration Project in Canada [82], the Urban-Route Buses Trial Project in China [83], the Brazilian Fuel Cell Bus Project in Brazil [84], and the Zero Emission Bay Area (ZEBA) demonstration in California, United States [85]. Fig. 14 shows the main components in a typical FCEB [86]. It is clear from the figure that the design utilizes the space flexibility on the roof, front, and back of the bus. Thus, the bus floor could be made lower compared to other conventional buses. The main components are very similar to those previously described for the L-FCEV, with PEMFCs and PAFCs being the most commonly-used types for the stack and high-voltage batteries used for regenerative braking energy recovery and a better dynamic response. However, the immaturity of fuel cell technologies and the lack of mass production and manufacturing are making FCEBs economically-uncompetitive with conventional buses and other competing novel technologies. This is more evident for FCEBs than L-FCEVs due to the higher durability and reliability requirements in FCEBs. Nevertheless, progress in the durability and cost of fuel cells designed for FCEBs is underway due to the increasing number of demonstration programs and the efforts by both fuel cell and bus developers such as Ballard, Hydrogenics, and Daimler AG.

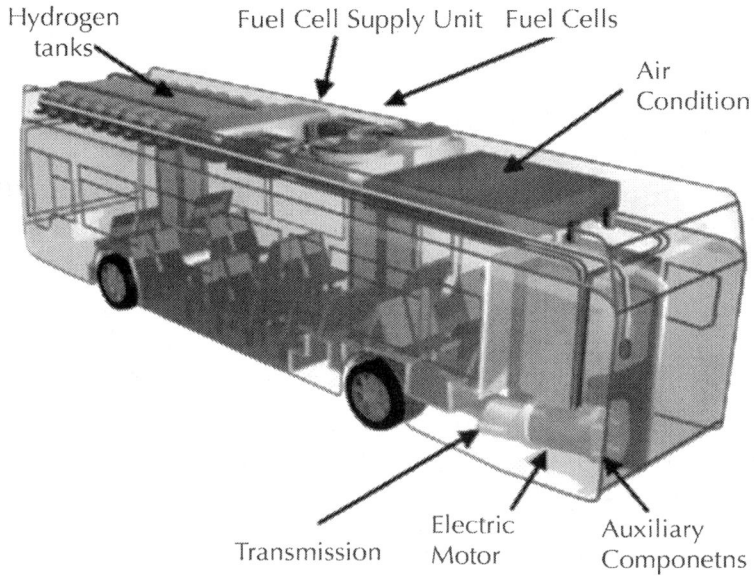

Figure 14: Main components in a typical FCEB based on the Mercedes-Benz Citaro Fuel Cell EcoBus [86].

Even though in the transportation sector, L-FCEVs, FCEBs, and fuel cell-based APUs are dominating the market, several impressive demonstrations and milestones have recently been achieved for H-FCEVs. Vision has announced the commercialization of what it claims to be the world's first hydrogen fuel cell-based heavy-duty class-8 truck. The 8.4 m long by 2.9 m high truck runs on a hybrid hydrogen fuel cell/Li-ion battery drive system capable of running at temperatures as low as −26 °C and as high as 43 °C while providing peak power of about 400 kW and a top speed of about 105 kmph at total vehicle weight of more than 36,000 kg. The fuel cell system (developed by Hydrogenics) has an output of 65 kW, a range of more than 320 km, and refueling time between

4 and 7 min. A purchase order for 100 of Vision›s trucks has been signed by the Total Transportation Services (TTS) with a total worth of $27 million [87]. Similarly, Heliocentris is developing a hybrid waste disposal heavy-duty truck that uses a hybrid diesel engine/ fuel cell system in Germany. The diesel engine will be used for propulsion while the fuel cell will be used for waste collection, management, and disposal. While for rail vehicles, Guo et al. [88] designed and simulated the performance of a proposed power system for a hybrid switcher locomotive that uses a SOFC power plant, lead-acid batteries, and ultracapacitors for the locomotive›s propulsion. The results of the study show that using the proposed control strategy for regulating the load sharing between the three sources would meet the locomotive›s required power demand with high efficiency. On the other hand, Miller et al. [89]found that using only a fuel cell for the power plant of a 1.2 MW yard switcher locomotive is actually more effective than using a hybrid fuel cell design coupled with auxiliary storage devices. This is because for a switcher locomotive, complex transient power requirements are not an issue since adhesion between the wheels and the rails is the limiting factor for tractive power, and not the peak power available. Additionally, potential for recovery of regenerative braking energy is low and the added complexity, volume, and weight of a hybrid design could be problematic.

Aerial Propulsion

Other applications for fuel cells include the space and aviation industries with small unmanned aerial vehicles (UAVs) being the main focus for fuel cells in the aerial propulsion sector. UAVs are mainly used for surveying, surveillance, and reconnaissance purposes due to their stealth nature and lack of risk to human life. And with the ever-increasing interest in UAVs by military authorities and commercial parties, the development of more durable and reliable propulsion systems is a necessity. Fuel cells (mostly PEMFCs with few SOFCs) are clearly becoming the ideal candidates for powering future UAVs. The stealth nature of UAVs is facilitated by fuel cells' static operation and low heat dissipation, two advantages over UAVs with internal combustion engines. And even though batteries share those two advantages with fuel cells, the low energy density and large weight of batteries make fuel cell UAVs superior to UAVs with batteries. The lighter weight and higher energy density of fuel cells allow for greater mission range and endurance (up to 24 h [64]) as compared to an average of one hour for battery UAVs. Additionally, the modularity of fuel cells makes them promising to use for small-scale applications such as UAVs, contrary to combustion engines that suffer from low efficiencies when designed for small-scale applications. Thus, about 20 UAVs with fuel cells have been demonstrated so far [64]. A comparison conducted by Bradley et al. [90] between five different potential small-scale UAV propulsion systems concluded that a PEMFC system with compressed gaseous hydrogen has the highest potential for both range and endurance. The other four systems were a propane-based SOFC, a lithium polymer battery, a zinc air battery, and a small internal combustion engine. The results of the comparison are summarized in Fig. 15.

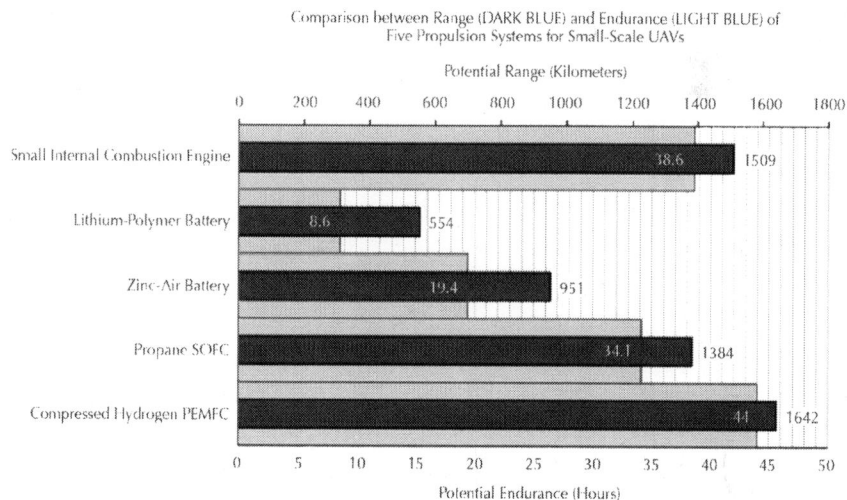

Figure 15: Comparison between different proposed UAV small-scale propulsion systems. Source: Data Sourced from [90].

Kim et al. [91] designed, built, and tested a UAV with a hybrid fuel cell/battery propulsion system with on-board hydrogen production using sodium borohydride. They used the fuel cell to meet the steady-state power demand during cruising while both the battery and the fuel cell provided the power required for the more power-demanding take-offs and maneuvers. They concluded that using fuel cell systems in UAVs is more efficient than battery- and combustion-based UAVs for long endurance flights. Another innovative and highly-efficient concept was NASA's Helios UAV that used a hybrid propulsion system consisting of photovoltaic cells, a regenerative fuel cell, and backup batteries for achievement of both high altitudes and long flights. The regenerative fuel cell functions as an electrolyzer that generates hydrogen during the day using power from the photovoltaic cells and then functions as a fuel cell that uses this hydrogen for flight during the night. Other demonstrations include EnergyOr Technologies demonstration of a long endurance flight for more than 10 h with its fuel cell-powered UAV using the company's own fuel cell system [29]. On the other hand, the market of manned military and commercial air vehicles is still impractical for fuel cells due to the market's high energy density, power density, durability, and reliability requirements. Nonetheless, in 2008 Boeing announced the successful testing of a small airplane using a hybrid hydrogen PEMFC system, developed by Intelligent Energy, coupled with an engine and Li-ion batteries [92]. A few other manned fuel cell-based airplanes have also been demonstrated afterwards, of which, a demonstration by DLR in Germany used a system that relied solely on a hydrogen fuel cell with a range of 750 km, an endurance of 5 h, and a fuel cell efficiency of 52% [93]. In Japan, a team of researchers successfully designed, simulated, and tested a fuel cell-powered high-altitude balloon [94]. Moreover, the space industry has been one of the first to adopt fuel cells. NASA used AFCs and PEMFCs for its manned space programs during the 1960s. However, during the past 10 years, interest in fuel cells for space applications has been revived [95], [96] and [97]. Fuel cells are attractive for space applications due to their many advantaged compared to other power generation technologies. However, the fact that water is a byproduct of the electrochemical reactions within a fuel cell makes it even more attractive for space applications where air, water, and food supplies are of the utmost importance.

Marine Propulsion

Even though the most common use for fuel cells in the marine industry is as APUs on-board of boats and yachts, as previously mentioned, promising future marine propulsion markets for fuel cells exist in submarines, ferries, underwater vehicles, boats, yachts, and even cargo ships. Fuel cells provide their regular benefits for ships and ferries, such as low emissions, high efficiency, and static operation. However, issues related to reliability, lifetime, shock resistance, and tolerance to the salt content of sea air are yet to be resolved. Currently, PEMFCs, SOFCs, and MCFCs hold the most potential for the marine fuel cell market. In 2003, the first yacht with a certified hybrid PEMFCs/lead-gel batteries system for both propulsion and APU was successfully demonstrated in Germany [98]. While for commercial ships, in 2008, the world's first commercial passenger ship running on a hybrid PEMFCs/lead-gel battery system was put into service in Germany. The ship has a capacity of 100 passengers and offers twice the efficiency of a conventional diesel-based ship [99]. Another innovative and self-sufficient hydrogen fuel cell boat was developed in Austria in 2009 [100]. The boat's propulsion system consists of photovoltaic panels, an electrolyzer, a high-pressure hydrogen storage system, and a fuel cell. The system uses solar energy to decompose water in the electrolyzer into hydrogen and oxygen. The hydrogen is then fed to the fuel cell system for propelling the boat. According to the developers, the boat has a range of 80 km, which is twice the range of a conventional battery-based boat. Furthermore, in 2011, the world's first hydrogen fuel cell-propelled ferry has been in daily operation in Germany [29]. Alkaner and Zhou [101] concluded that using conventional marine fuels reformation on-board of ships for hydrogen generation is the solution to overcome the low volumetric density of hydrogen and to make fuel cells a practical propulsion alternative in commercial ships. The researchers also concluded based on a life cycle assessment (including manufacturing, fuel production, operation, and decommissioning) that using MCFCs with on-board reformation holds no significant advantage over conventional diesel engines with respect to neither energy nor environment. This is mainly due to the immaturity of MCFC technology, the low durability of MCFC stacks (about 5 years), and the use of energy-intensive materials and manufacturing processes due to the lack of commercial production. Leo et al. [102] conducted an extensive exergy analysis of two proposed fuel configurations for power generation in marine applications. The first configuration used a PEMFC coupled with a liquid methanol reformation stage. The second configuration used a DMFC. The results of the analysis showed that both systems had similar exergy efficiencies (even though more exergy was lost in the DMFC configuration) and for more conclusive results, a thermoeconomic analysis should be conducted in order to take the economic factor into account. Submarines that use oxygen/hydrogen fuel cells for propulsion and auxiliary load requirements have also been successfully deployed. A series of PEMFC-based hybrid submarines have been developed in Germany for the German and Italian navies [103]. The submarines use a hybrid power system consisting of a Siemens fuel cell system, diesel generator, and a high-voltage battery. Some remarkable advantages of fuel cell-based submarines are that they can stay underwater without surfacing for refueling much longer when compared to a conventional diesel electric submarine (the German submarines are able to stay underwater for weeks fueled by the stored oxygen and hydrogen in comparison to a limit of only two days if the propulsion system was purely diesel-based), have efficiencies as high as 70% due to the usage of pure oxygen as the oxidant in the fuel cell instead of air, have lower heat and magnetic signatures when compared to nuclear- or diesel-based submarines, and produce almost no noise due to the static operation of the fuel cell. These advantages make fuel cell-based submarines highly undetectable and ideal for modern military mission requirements, which encouraged the Italian, Greek, and South Korean navies to invest and take interest in fuel cell submarines as well [104].

Fig. 16 summarizes the main applications in the portable, stationary, and transportation sectors as per the above discussions. Table 9 lists the major fuel cell system solutions developers in the world with the types of fuel cell systems they develop, the company's country, and the targeted markets of their fuel cell systems solutions.

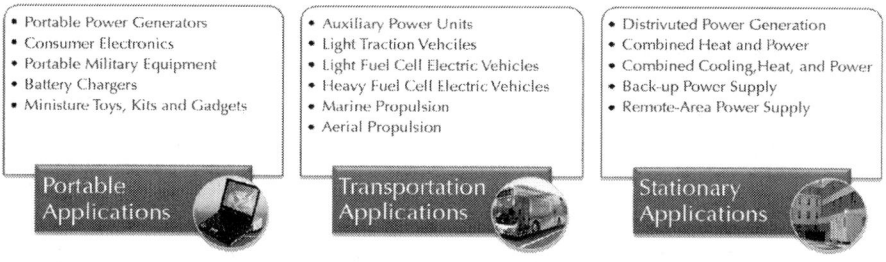

Figure 16: Summary of fuel cell applications.

Table 9: Major fuel cell system solutions development companies

Country	#	Company	FC type(s)	Market(s)
United States	1	Acumentrics	SOFCs	• RAPS • Portable military equipment • Industrial and residential distributed CHP generation • Commercial distributed power generation • EPS
	2	Altergy	PEMFCs	• EPS
	3	Bloom Energy	SOFCs	• Commercial distributed power generation • EPS
	4	Boeing	PEMFCs	• Aerial propulsion
				• APUs

5	ClearEdge Power[a]	PEMFCs PAFCs	• Residential and com-mercial distributed power generation • Residential and com-mercial distributed CHP generation • EPS • APUs • L-FCEVs • H-FCEVs
6	Delphi	SOFCs	• APUs
7	EnerFuel	PEMFCs	• Residential and com-mercial distributed power generation • Residential and com-mercial distributed CHP generation • EPS • APUs
8	First Element	PEMFCs	• EPS • Industrial and commer-cial distributed power generation • RAPS
9	Ford	PEMFCs	• L-FCEVs
10	FueCell Energy[b]	MCFCs SOFCs	• Commercial and indus-trial distributed power generation • Commercial and indus-trial distributed CHP generation
11	General Mo-tors	PEMFCs	• L-FCEVs • H-FCEVs
12	Infinity	PEMFCs	• APUs
13	Infintium	PEMFCs	• Material handling
14	Microcell	PEMFCs	• EPS • Portable power generators • Commercial distributed CHP generation • Commercial distributed CCHP generation
15	Motorola	DMFCs	• Consumer electronics

				• EPS
	16	MTI Micro	DMFCs	• Consumer electronics • Battery chargers
	17	Neah Power	DMFCs	• Consumer electronics • Portable power generators • Portable military equip-ment
	18	Nuvera	PEMFCs	• Material handling • APUs • L-FCEVs • Residential distributed power generation
	19	Oorja	DMFCs	• Material handling
	20	Plug Power	PEMFCs	• Material handling
	21	Protonex	PEMFCs SOFCs	• Portable military equip-ment • UAVs • Battery chargers • APUs • Portable power generators • EPS • RAPS
	22	ReliOn	PEMFCs	• EPS • RAPS
	23	Ultra Elec-tronics AMI	SOFCs	• Consumer electronics • APUs • Battery chargers • Portable military equip-ment • Portable power generators
	24	UltraCell	RMFCs	• Consumer electronics • Portable power generators • Portable military equip-ment
	25	Vision	PEMFCs	• H-FCEVs
Japan	1	Canon	PEMFCs	• Consumer electronics

	2	Fuji Electric	PAFCs PEMFCs	• Industrial and commercial distributed power generation • Industrial and commercial CHP generation
	3	Hitachi	SOFCs DMFCs	• Residential distributed CHP generation • Consumer electronics • Portable power generators
	4	Honda	PEMFCs	• LTVs • Residential distributed CHP generation • L-FCEVs
	5	IHI	PEMFCs MCFCs	• Residential and commercial distributed power generation • APUs
	6	Mitsubishi	PEMFCs SOFCs	• L-FCEVs • H-FCEVs • Residential and commercial distributed power generation • Residential and commercial distributed CHP generation • Marine propulsion
	7	NEC	DMFCs	• Consumer electronics
	8	Nissan	PEMFCs DMFCs	• L-FCEVs
	9	Panasonic[c]	PEMFCs DMFCs	• Residential distributed CHP generation • Portable power generators • Consumer electronics
	10	Sony	Microbial FCs DMFCs	• Consumer electronics • Battery chargers
	11	Suzuki	PEMFCs	• LTVs • L-FCEVs
	12	Toshiba	DMFCs PEMFCs PAFCs	• Consumer electronics • Battery chargers • Residential distributed CHP generation • EPS

	13	Toyota	PEMFCs	• H-FCEVs
			SOFCs	• L-FCEVs • Material Handling • Residential distributed CHP generation
	14	Yamaha	DMFCs	• LTVs
Germany	1	Baxi Inno-tech	PEMFCs	• Residential distributed CHP generation
	2	BMW	PEMFCs	• APUs • L-FCEVs
	3	Daimler[d]	PEMFCs	• L-FCEVs • H-FCEVs
	4	FutureE	PEMFCs	• EPS • RAPS • Distributed power generation
	5	Heliocentris	PEMFCs	• Toys and educational kits • RAPS • EPS
	6	Proton Motor	PEMFCs	• H-FCEVs • Material handling • EPS • Marine propulsion
	7	Schunk	PEMFCs	• Battery chargers • General-purpose stacks and systems
	8	SFC Energy	DMFCs	• Battery chargers • RAPS • EPS • Portable power generators • Portable military equipment
	9	Siemens	DMFCs PEMFCs SOFCs	• Consumer electronics • Marine propulsion • Industrial distributed CHP generation
	10	Volkswagen[e]	PEMFCs	• L-FCEVs • H-FCEVs • APUs

Canada	1	AFCC	PEMFCs	• L-FCEVs • H-FCEVs
	2	Ballard[f]	PEMFCs DMFCs	• EPS • Commercial distributed CHP generation • RAPS • Material handling • H-FCEVs
	3	DDI Energy	SOFCs	• RAPS • EPS • Residential and commercial distributed power generation • Residential and commercial distributed CHP generation • APUs
	4	Hydrogenics	PEMFCs	• RAPS • Material Handling • H-FCEVs • Marine propulsion • Aerial propulsion • EPS • L-FCEVs
	5	New Flyer	PEMFCs	• H-FCEVs
	6	Palcan	PEMFCs	• EPS • RAPS • Portable power generators
United Kingdom	1	AFC Energy	AFCs	• Industrial distributed power generation
	2	Ceres Power	SOFCs	• Residential distributed CHP generation • APUs • Portable power generators • EPS

	3	Intelligent Energy	PEMFCs	• L-FCEVs • LTVs • EPS • Residential and commercial distributed CHP generation • Consumer electronics
	4	Morgan	PEMFCs	• L-FCEVs
	5	Riversimple	PEMFCs	• L-FCEVs
South Korea	1	Hyundai	PEMFCs	• H-FCEVs • L-FCEVs
	2	Kia	PEMFCs	• L-FCEVs
	3	LG[g]	DMFCs SOFCs	• Consumer electronics • Industrial and commercial distributed CHP generation
	4	Samsung	DMFCs PEMFCs SOFCs	• Consumer electronics • Portable power generators • Portable military equipment • Distributed power generation
Sweden	1	Cellkraft	PEMFCs	• RAPS • EPS • Portable military equipment
	2	myFC	PEMFCs SOFCs	• Consumer electronics • Battery chargers
	3	Powercell	PEMFCs	• APUs • EPS
Taiwan	1	Antig	DMFCs	• Consumer electronics • Battery chargers • Portable power generators
	2	APFCT	PEMFCs	• LTVs
	3	M-Field	PEMFCs	• EPS • APUs

Denmark	1	H2 Logic	PEMFCs	• Material handling
	2	Serenergy	PEMFCs RMFCs	• EPS • APUs • Material handling • L-FCEVs • Battery chargers • Portable power generators
	3	Topsoe Fuel Cell	SOFCs	• Residential distributed CHP generation
France	1	BIC[h]	N/A	• Consumer electronics • Battery chargers
	2	Peugeot	PEMFCs	• APUs • L-FCEVs
	3	Renault	PEMFCs	• L-FCEVs
Switzer-land	1	Hexis	SOFCs	• Residential distributed CHP generation
	2	MES	PEMFCs	• L-FCEVs • EPS • Aerial propulsion • UAVs • LTVs • Portable power generators
Italy	1	SOFCpower	SOFCs	• Residential and commercial distributed CHP generation • RAPS
	2	Fiat[i]	PEMFCs	• L-FCEVs • H-FCEVs
Belgium	1	Van Hool	PEMFCs	• H-FCEVs
Finland	1	Convion[j]	SOFCs	• Residential distributed CHP generation
Estonia	1	Elcogen	SOFCs	• General-purpose stacks

Nether-lands	1	Nedstack	PEMFCs	• EPS • RAPS • Material handling • H-FCEVs • Industrial and residential distributed CHP generation • Marine propulsion • LTVs • L-FCEVs
Australia	1	Ceramic Fuel Cells	SOFCs	• Residential and commercial distributed CHP generation • Residential and commercial distributed power generation
Singa-pore	1	Horizon	PEMFCs DMFCs	• UAVs • Consumer electronics • Battery chargers • Portable power generators • Toys and educational kits • RAPS • EPS • L-FCEVs
Greece	1	Tropical	PEMFCs	• H-FCEVs • Residential and commercial distributed CHP generation • Portable power generators • L-FCEVs • LTVs • EPS • RAPS

[a]On February 2013, ClearEdge Power acquired UTC Power.

[b]And its subsidiary Versa.

[c]And its subsidiary Sanyo.

[d]And its brand Mercedes-Benz.

[e]And its brands Audi and Skoda.

[f]On July 2012, Ballard acquired IdaTech.

[g]In June 2012, LG acquired Rolls-Royce Fuel Cell Systems.

[h]On November 2011, BiC acquired Angstrom Power.

iAnd its brand Chrysler.

jOn January 2013, Wärtsilä fuel cell activities have been moved to a new company—Convion.

CURRENT AND FUTURE R&D

Current Status

Indeed, the fuel cell industry has gone though many milestones and accomplishments in the past decade. For instance, the cost of fuel cells customized for fuel cell electric vehicles, projected to a production level of 500,000 units per annum, went through a cost reduction of 83% in the span of 2002–2012, from $275/kW to $47/kW [9]. This value jumps to $219/kW if projected to a production level of only 1000 units per annum [105], emphasizing the impact mass production has on cost. One of the main reasons for this cost reduction was due to the continuous decrease in the platinum group metals (PGM) loading in PEM stacks. PGM-loading decreased by two orders of magnitude since the 1960s due to major development efforts in electrocatalyst fabrication and nanotechnology [106]. Patents granted in different alternative energy sectors between the years 2002 and 2012 were led in each single year by the fuel cell industry [107], as seen from Fig. 17. 44% of the patents in the fuel cell sector between 2002 and 2012 went to developers from the United States, followed by 33% in Japan, 7% in Korea, and 6% in Germany; with General Motors, Honda, Toyota, Samsung, and UTC Power securing more than 60% of these patents, in that order. The number of patents granted is a reflection of the level of industrial research taking place in the different renewable sectors. Table 10 lists some of the most important achievements fuel cell R&D went through in the year 2012 in the United States [105].

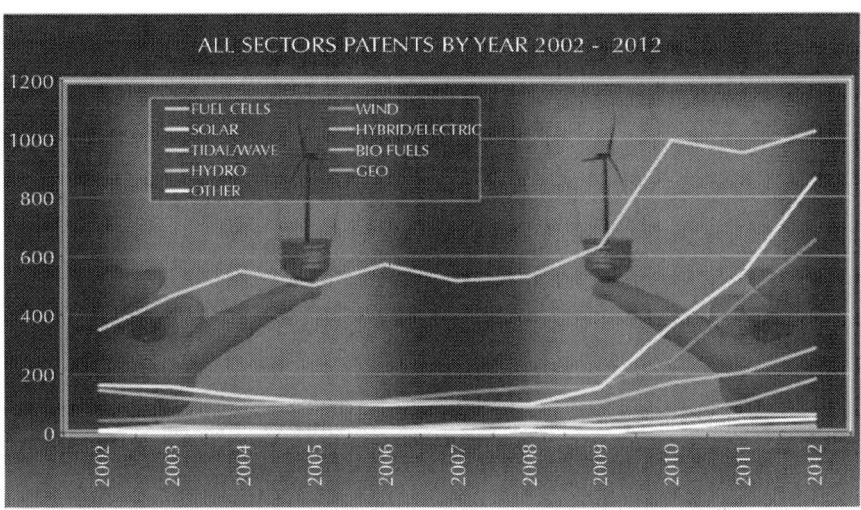

Figure 17: Patents granted in the alternative energy sectors between 2002 and 2012 [107].

Table 10: Major fuel cell R&D milestones in the United States in 2012 (adapted from [105])

R&D focus	2012 Milestone(s)
FC system costa	A 4% cost reduction from 2011 ($49 to $47/kW) due to platinum-loading reduction, cell power density increase, ejector system modifications, improved system controller, radiator size reduction
FC system durabilityb	4000 h in laboratory testing was achieved (100% increase from 2006) due to modified anode and cathode catalysts with highly-active and durable oxygen evolution properties
Catalyst	• Development of dealloyed catalysts with high mass activity, high durability, and high cell voltage under high current density • Reduction of platinum-loading by 15% for nanostructured thin film (NSTF) catalysts to 0.14–0.18 g/kW

Portable FC	• Power density of direct dimethyl ether fuel cell improved by 60% from 2011, which approaches direct methanol fuel cells at low currents, due to a new anode catalyst • Mass activity of direct methanol fuel cell improved by 150% due to a new anode catalyst
BoP components	A novel composite membrane humidifier is projected to meet the target of $100 under mass production

[a]Projected at a production level of 500,000 units per annum.

[b]Defined as the time it takes the stack to lose 10% of its original voltage.

Future Targets

Nonetheless, the industry still has a long way ahead in order to achieve the long-awaited goal of widespread commercialization. It is important to realize that advancements in fuel cell technology go hand-in-hand with advancements in hydrogen production, storage, and delivery technologies. With this relation in mind, we are still in-need for fundamental breakthroughs in material engineering, nanotechnology, transport phenomena, electrocatalysts engineering, stack engineering, measurement technologies, simulation of molecular processes, auxiliary components development, and multi-phase science in order to decrease the cost and increase the durability of fuel cells to meet future targets. For example, fundamental knowledge of liquid water formulation and interactions can ensure efficient water balance and avoid flow misdistribution which would enhance the fuel cell's performance and efficiency [28]. The key limitations facing the fuel cell industry that call for an R&D focus from both the industrial and academic communities include:

- identifying, modeling, and mitigating MEA degradation mechanisms;
- developing electrolyte materials that maintain conductivity and stability over a wide range of temperature and humidity;

- minimizing/eliminating catalysts PGM loading;
- maximizing membrane and catalyst impurities tolerance;
- identifying membrane and catalyst stability with voltage and humidity cycling;
- developing bipolar plate designs with reduced pressure drops, volume, and weight;
- developing durable seals, bipolar plates, and MEAs for high-temperature operation;
- developing air management techniques with reduced noise, cost, and parasitic loads;
- developing water management techniques capable of handling a wide range of operation conditions based on a better understanding of water transport phenomena;
- improving and simplifying fuel reformation methods;
- improving and unifying standards for accelerated testing; and
- conducting cost analyses for niche markets along with updated status reports.

Table 11 details the main targets for the ongoing fuel cell research and development efforts in the United States with respect to both individual components in a fuel cell stack and components in the fuel cell system. Table 12 summarizes the set future targets for fuel cell technology in the United States with respect to the different fuel cell markets in order for mass production and widespread commercialization of fuel cells to become practical goals instead of optimistic speculations.

Table 11: Fuel cell R&D targets (adapted from [9] and [108])

R&D focus	R&D targets
Electrolytes	<ul><li>Develop electrolyte material (polymer, phosphoric/ solid acid, solid oxide, molten carbonate, anion-exchange) with improved conductivity over temperature and humidity ranges</li><li>Develop electrolyte material (polymer, phosphoric/ solid acid, solid oxide, molten carbonate, anion-exchange) with improved mechanical, thermal, and chemical stability over temperature and humidity ranges</li><li>Develop electrolyte material (polymer, phosphoric/ solid acid, solid oxide, molten carbonate, anion-exchange) with reduced (or eliminated) fuel crossover over temperature and humidity ranges</li><li>Design ionomer membranes using scalable fabrication processes</li><li>Design ionomer membranes with reduced cost</li><li>Design ionomer membranes with improved mechanical, thermal, and chemical stability over temperature and humidity ranges</li><li>Evaluate membrane tolerance of air, fuel, and system impurities</li><li>Evaluate membrane mechanical stability with relative humidity cycling</li><li>Identify electrolyte mechanical and chemical degradation mechanisms</li><li>Develop mitigation strategies for electrolyte mechanical and chemical degradation mechanisms</li></ul>

Catalysts	• Develop catalysts with reduced (or eliminated) precious metal loading for low-, medium-, and high-temperature fuel cells • Develop catalysts with improved specific and mass activities • Develop catalysts with improved stability with potential cycling • Develop catalysts with tolerance of air, fuel, and system impurities • Develop non-PGM catalyst for PEM and anion-exchange membrane fuel cells • Increase catalysts utilization • Develop high-temperature fuel cell catalysts with improved activity and durability • Reduce corrosion of catalyst supports • Develop catalyst support material and structure with reduced cost • Develop catalyst supports with improved loading and thickness of non-PGM catalyst • Optimize catalyst/support interactions and microstructure • Develop anode catalyst for non-hydrogen fuel cells
GDLs	• Improve GDL pore structure, morphology, and physical properties • Improve GDL coating for better water management and more stable operation • Develop GDL material and structures with reduced area-specific resistance • Identify GDL corrosion and degradation mechanisms
MEAs and unit cells	• Optimize catalyst/support/ionomer/membrane mechanical and chemical interactions • Minimize MEA interfacial resistance • Integrate membranes, catalysts, and GDLs into unified MEAs • Integrate catalysts, supports, and electrolytes for high-temperature fuel cells • Address MEA freezing and thawing issues • Expand MEA operation temperature and humidity ranges • Improve MEA and cell stability under voltage and humidity cycling • Develop mitigation strategies for the effect of air, fuel, and system impurities • Characterize and test MEAs and cells before, during, and after fabrication and operation
Seals (gaskets)	• Develop seals for high-temperature fuel cells

Bipolar plates and interconnects	• Develop interconnects for high-temperature fuel cells • Develop electrolyte reservoir plates for phosphoric acid fuel cells • Decrease weight and volume of bipolar plates • Develop coatings for bipolar plates that eliminate corrosion • Develop bipolar plate materials and coatings for reduce cost • Identify bipolar plates mechanical and chemical degradation mechanisms • Develop mitigation strategies for bipolar plates mechanical and chemical degradation mechanisms
Stack-level operation	• Model stack impurities effects • Model stack durability and degradation • Model stack freezing and thawing effects • Model performance of upgraded stack components • Identify stack long-term failure modes using experimentation • Model stack mass transport and validate using experimental data • Optimize stack water management
BoP components	• Reduce cost of chemical and temperature sensors in stationary applications • Improve reliability and durability of chemical and temperature sensors in stationary applications • Meet packaging, cost, and performance requirements for air management mechanisms in stationary and transportation applications • Minimize parasitic load of air management mechanisms in stationary and transportation applications • Reduce noise level of air management mechanisms in stationary applications • Develop non-toxic coolants with low electrical conductivities • Increase efficiency, durability, and reliability of humidifiers in transportation applications • Develop humidifier materials and new humidification concepts for transportation applications • Minimize parasitic loads of humidifiers in transportation applications • Develop lightweight humidifier material with reduced cost in transportation applications

Fuel reformation	• Develop fuel-flexible reformers • Develop reformation catalysts and hardware that generate hydrogen-rich gas • Minimize fuel reformation cost • Improve reformers tolerance to impurities • Develop gas clean-up with low cost • Integrate subsystems of fuel reformers • Integrate thermal loads of fuel reformers • Eliminate hardware, piping, sensors, and controls from reactor
System-level operation	• Model system impurities effects • Model system durability and degradation • Minimize carbon dioxide migration in alkaline fuel cells • Improve start-up time and stability for high-temperature fuel cells
Performance characterization	• Perform cost analyses for automotive and bus applications • Perform cost analyses for emerging applications including auxiliary power units (APUs), emergency back-up systems, and material handling • Annually update technology status • Conduct tradeoff analysis between rated power and efficiency • Conduct tradeoff analysis between start-up energy and start-up time • Conduct tradeoff analysis between hydrogen quality and fuel cell performance and durability
Experimental testing and diagnostics	• Determine long-term stack failure modes using experimental methods • Determine system emissions using experimental methods • Characterize component and stack properties before, during, and after operation using experimental methods • Develop accelerated-testing mechanisms for durability in stationary applications

Table 12: Fuel Cell Commercialization Targets (Adapted from [108] and [109])

Market	Characteristics	Unit	Current Status	Future Target

80 kWe Automotive Transportation[a]	Electric Efficiency[b]	%	59	60
	Power Density	$W\text{-}L^{-1}$	400	850
	Specific Power	$W\text{-}kg^{-1}$	400	650
	Cost[c]	$\$\text{-}kW_e^{-1}$	49	30
	Cold Start-Up Time at −20 °C[d]	s	20	30
	Cold Start-Up Energy at -20 °C[e]	MJ	7.5	5
	Cold Start-Up Time at 20 °C[d]	s	<10	5
	Cold Start-Up Energy at 20 °C[e]	MJ	N/A	1
	Durability[f]	h	2,500	5,000
1–10 kWe Small Residential CHP[g]	Electric Efficiency[h]	%	30–40	>45
	CHP Efficiency[i]	%	80–90	90
	Cost[j]	$\$\text{-}kW_e^{-1}$	2,300–4,000	1,500
	Dynamic Response Time[k]	min	5	2
	Cold Start-Up Time at 20 °C	min	<30	20
	Degradation Rate[l]	$\dfrac{powerloss\%}{1.000hours}$	2	0.3
	Durability[m]	h	12,000	60,000
	Availability[n]	%	97	99
100–3,000 kWe Medium CHP[o]	Electric Efficiency[h]	%	42–47	>50
	CHP Efficiency[i]	%	70–90	90
	Equipment Cost for Natural Gas Fuel[p]	$\$\text{-}kW_e^{-1}$	2,500–4,500	1,000
	Equipment cost for Biogas Fuel[p]	$\$\text{-}kW_e^{-1}$	4,500–6,500	1,400
	Number of Lifetime Outages[q]	–	50	40
	Durability[r]	h	40,000–80,000	80,000
	Availability[n]	%	95	99

<2 W Micro Portable[s]	Specific Power	W-kg^{-1}	5	10
	Power Density	W-L^{-1}	7	13
	Specific Energy	Wh-kg^{-1}	110	230
	Energy Density	Wh-L^{-1}	150	300
	Cost[t]	$-system^{-1}	150	70
	Durability[u]	h	1,500	5,000
	MTBF[v]	h	500	5,000
10–50 W Small Portable[s]	Specific Power	W-kg^{-1}	15	45
	Power Density	W-L^{-1}	20	55
	Specific Energy	Wh-kg^{-1}	150	650
	Energy Density	Wh-L^{-1}	200	800
	Cost[w]	$-system^{-1}	15	7
	Durability[u]	h	1,500	5,000
	MTBF[v]	h	500	5,000
100–250 W Medium Portable[s]	Specific Power	W-kg^{-1}	25	50
	Power Density	W-L^{-1}	30	70
	Specific Energy	Wh-kg^{-1}	250	640
	Energy Density	Wh-L^{-1}	300	900
	Cost[x]	$-system^{-1}	15	5
	Durability[u]	h	2,000	5,000
	MTBF[v]	h	500	5,000
1–10 kWe APU[ag]	Electric Efficiency[y]	%	25	40
	Power Density	W-L^{-1}	17	40
	Specific Power	W-kg^{-1}	20	45
	Factory Cost[z]	$-kW^{-1}	2,000	1,000
	Dynamic Response Time[k]	min	5	2
	Cold Start-Up Time at 20 °C	min	50	30
	Standby Start-Up Time	min	50	5
	Degradation Rate[aa]	$\frac{powerloss\%}{1.000hours}$	2.6	1
	Durability[ab]	h	3,000	20,000
	Availability[ac]	%	97	99

Transit Buses[ad]	Equivalent Mileage	miles per gallons diesel-equivalent	7	8
	Maintenance Cost[q]	$-mile^{-1}	1.20	0.40
	Operation Time	h-week^{-1}	133	140
	Power System[ae] Cost[af]	$	700,000	200,000
	Bus Cost[af]	$	2,000,000	600,000
	Availability	%	60	90
	Power System Durability	h	12,000	25,000
	Bus Durability	years	5	12

[a]For a direct-hydrogen PEM fuel cell system excluding hydrogen storage, power electronics, and the electric drive with current statues in 2011 and future targets set in 2020.

[b]Defined as $\dfrac{DC\,output\,energy}{LHV\,of\,hydrogen}$ at 25% rated power (which corresponds to the peak efficiency).

[c]Projected at a production level of 500,000 units per annum.

[d]To 50% rated power.

[e]Defined as the energy consumed from cold start to 50% rated power based on LHV of hydrogen.

[f]Defined as the time it takes the stack to lose 10% of its original voltage.

[g]Fuel cell system operating on natural gas delivered through pipelines at typical distribution pressure with current statues in 2011 and future targets set in 2020.

[h]Defined as $\dfrac{regulated\,AC\,output\,energy}{LHV\,of\,fuel}$.

[i]Defined as

$$\dfrac{regulated\,AC\,output\,energy + useful\,recovered\,thermal\,energy}{LHV\,of\,fuel}$$

where domestic space and water heating are the useful applications for the recovered thermal energy.

jProjected at a production level of 500,000 units per annum for a system with an average AC output of 5 kW_e while the system is running where cost includes all CHP system components and equipment (with taxes and markup).

kDefined as the time it takes the system to respond to a 10% to 90% rated power demand change.

lWhere transient operation effects are included in the degradation tests.

mDefined as the time it takes the system to lose >20% of its original net power.

nDefined as the percentage of time the system is available for realistic operation (system unavailability due to maintenance etc.).

oFuel cell system, including fuel processor and auxiliary equipment, operating on natural gas delivered through pipelines at typical distribution pressure (current status vary according to the used technology) with current statues in 2011 and future targets set in 2020.

pWhere current cost is for the current ~30 MW per year production rate and future target cost is projected to ~100 MW per year both without installation costs.

qPlanned and forced.

rDefined as the time it takes the system to lose >10% of its original net power.

sFuel cell system (technology and fuel neutral) with current statues in 2011 and future targets set in 2020.

tProjected at a production level of 50,000 units per annum with installation costs included.

uDefined as the time it takes the system to lose 20% (somewhat variable with application) of its original net power where transient operation effects and offline degradation are included in the durability tests (tests are application-specific).

vMean Time Between Failures (MTBF) due to failure of any system component where transient operation effects and offline degradation are included in tests.

[w]Projected at a production level of 25,000 units per annum with installation costs included.

[x]Projected at a production level of 10,000 units per annum with installation costs included.

[y]Defined as $\dfrac{\text{regulated DC output energy}}{\text{LHV of fuel}}$.

[z]Projected at a production level of 50,000 units per annum for a system with an output of 5 kW where the cost includes material and labor to produce a complete system.

[aa]Where considerations for daily standby cycles, weekly shutdown cycles, exposure to vibrations, and variable operation condition are included in the degradation tests.

[ab]Defined as the time it takes the system to lose >20% of its original net power where considerations for daily standby cycles, weekly shutdown cycles, exposure to vibrations, and variable operation condition are included in the durability tests.

[ac]Defined as the percentage of time the system is available for realistic operation (system unavailability due to scheduled maintenance non-applicable).

[ad]Fuel cell (including auxiliary systems, power electronics, and hydrogen storage) and battery hybrid system with current status in 2012 and future targets set for the commercialization stage.

[ae]Defined for both the fuel cell system and the battery system.

[af]Projected at a production level of 700 units per annum.

[ag]Fuel cell system running on ultra-low sulfur diesel fuel with current statues in 2011 and future targets set in 2020.

FUEL CELL DESIGN LEVELS: THE UNIT CELL, THE STACK, AND THE SYSTEM

The unit cell is the heart of a fuel cell system where the basic electrochemical reactions take place. The building blocks of a single unit cell are called the membrane electrode assemblies (MEAs). An MEA consists of the polymer membrane in-between two electrodes and two gas diffusion layers (GDLs), also known as the porous transport layers (PTLs) and the gas diffusion media (GDM). Each electrode is a thin electrocatalyst layer (usually platinum deposited on the surface of carbon-supported powder) attached to either the membrane or the gas diffusion layer, as shown in Fig. 18[110]. This microscopic catalyst electrode layer is where the fuel cell's electrochemical reactions take place. In this layer, the reactant gases coming from the GDL interact electrically and ionically with the electrolyte membrane with the help of the electrocatalyst electrodes.

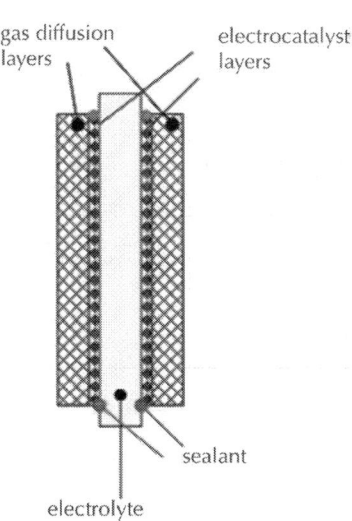

Figure 18: Membrane electrode assembly [110].

The potential of a single unit cell is typically between 0.5 and 0.8 V, which is too small for most practical applications. Thus, several unit cells are connected in-series to form what is known as a fuel cell stack, as shown in Fig. 19[111]. A fuel cell stack is significantly more complex than a single unit cell due to the requirements for current collection, thermal management, water management, humidification of gases, cell and gas separation, structural support, and oxidant and fuel distribution. In addition to the MEA, gas diffusion layers, heating and cooling plates, current collectors, end plates, clamping bolts, gaskets, insulators, and bipolar flow field plates are added to a fuel cell stack to satisfy these requirements, as shown in Fig. 20[112]. The specific roles of each of these components in a PEMFC stack are described and listed in Table 13. Along with materials engineering, the variety of options and strategies for stack design and configuration has made stack engineering one of the most critical and challenging aspects for the successful commercialization of fuel cells.

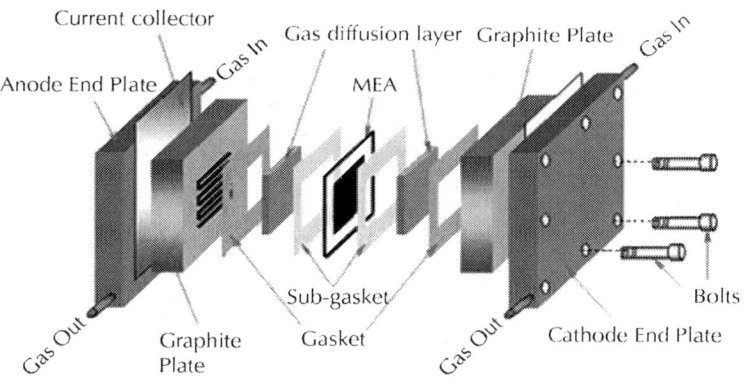

Figure 19: Single unit cell and a stack of cells [111].

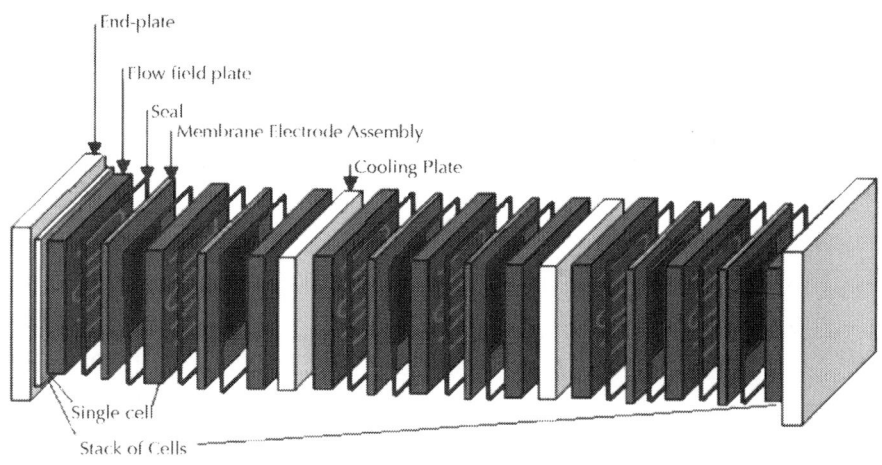

Figure 20: Main components in a fuel cell stack [112].

Table 13: PEMFC stack components and their functions

Component	Function
MEA	
Proton exchange membrane (electrolyte)	• Enables the protons to travel from the anode to the cathode. • A film barrier between the oxidation and reduction half reactions.
Electrocatalyst (electrode)	• Stimulates the oxidation and reduction reactions.
Gas diffusion layer	• Allows direct and uniform diffusion of hydrogen and oxygen to the catalyst layer (electrode). • Allows conduction of electrons to and from the catalyst layer. • Allows water formed at the cathode layer to exit. • Allows heat generated from the electrochemical reactions in the catalyst layer to exit. • Provides structural support to the "flimsy" MEA.

Flow field (bipolar) plate	• Channels oxygen and hydrogen to the electrodes via flow channels. • Channels water and heat away from the fuel cell. • Collects and conducts electrical current in-series. • Separates the gases in adjacent cells. • Forms the inner supporting structure of the fuel cell stack.
Gasket	• Aids in keeping the reactant gases in their respective regions in each cell.
Current collector	• Collects the current and passes it to/from the external circuit.
End plate	• Provides sufficient contact pressure in the stack to prevent leaking of reactants and to minimize the contact resistance between different layers.
Heating and cooling plates and manifold	• When the stack size is large, heating and cooling plates are internally used (between every 2–4 cells) to keep the stack's temperature near its optimum operating temperature.
Reactant gases manifold	• Feeds, externally or internally, each cell in the stack with oxygen and fuel.

A complete fuel cell system consists of the fuel cell stack in addition to the BoP subsystems. BoP subsystems are complementary components that provide the oxidant and fuel supply and storage, thermal management, water management, power conditioning, and instrumentation and control of the fuel cell system. The specific roles of each of these subsystems are described and listed in Table 14. A complete hydrogen-air PEMFC system is shown in Fig. 21[113]. Usually, the complexity of the overall fuel cell system increases with increasing fuel cell stack size as the temperature, pressure, water, and heat become more problematic and demanding.

Table 14: PEMFC balance-of-plant subsystems

Subsystem	Function
Water management	• Ensures all parts of the fuel cell are sufficiently hydrated without flooding. • Humidifies the incoming gases (especially to the anode). • Ensures proper water removal from the cathode. • Employs purge cycles and back pressure regulators for the removal of accumulated liquid water from the anode.
Thermal management	• Uses fans for active air cooling. • Uses pumps for circulation of cooling liquid through cooling plates. • Provides start-up heating in cold climates if required.
Gases management	• Employs an appropriate storage mechanism for hydrogen storage with pressure-reducing regulators. • Uses a fuel cell reformer in case of using hydrocarbons as hydrogen sources. • Employs a pump for hydrogen recirculation. • Employs a fan, blower, or compressor for air supply.
Power conditioning	• Converts the variable low-DC voltage output to usable DC power via a step-up DC–DC converter when required. • Inverts the variable low-DC voltage output to usable AC power via a switch-mode DC–AC inverter when required. • Employs a battery or an ultracapacitor to meet the power spike transients.

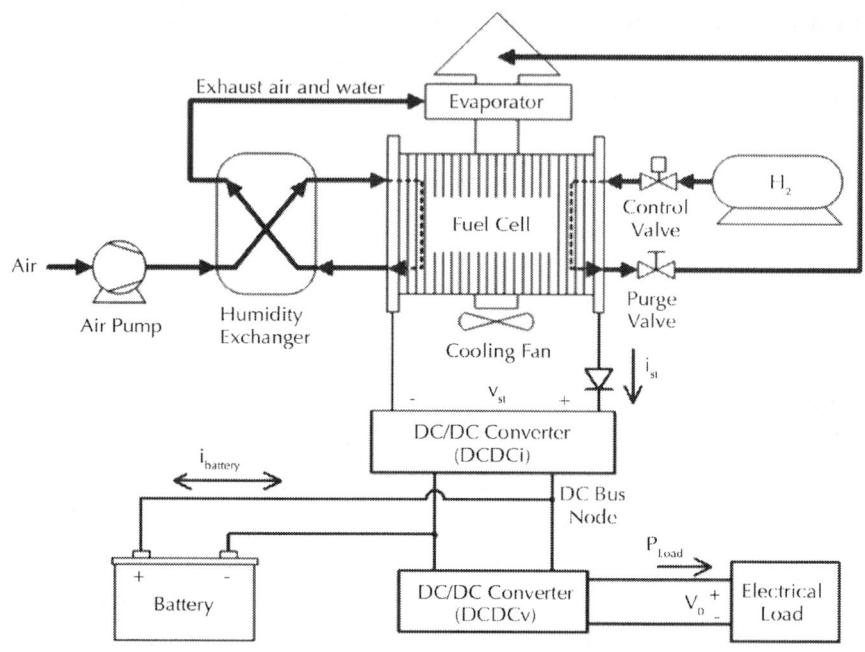

Figure 21: A complete hydrogen-air PEM fuel cell system [113].

THERMODYNAMIC AND ELECTROCHEMICAL PRINCIPLES OF HYDROGEN FUEL CELLS

In hydrogen fuel cells, the electrolyte conducts the H^+ ions from the anode to the cathode. The electrochemical reactions in a PEMFC occur simultaneously at the interfaces between the two catalyst layers and the membrane, as per Eqs. (1) and (2). Thus, H^+ passes through the proton-conductive acidic membrane from the anode to the cathode in response to the reactive attraction of hydrogen to oxygen, while e^- travels through an external circuit from the anode to be consumed at the cathodic reaction. The current of the e^- traveling through the external circuit gives us useful electrical work. At the cathode, $_{O2}$from the cathodic flow, H^+ that passed through the membrane, and e⁻ that passed through the external circuit all combine to form water. This water could be partially or completely removed with the cathodic outlet flow and could be in liquid or vapor form; depending on many interrelated factors such as the operation temperature, stoichiometric ratio of reactants, and design of the flow fields. Water accumulation at the cathode usually requires proper water and thermal management in order for the stack's performance not to decline. Thus, the overall electrochemical exothermic reaction is as per Eq. (3).

Reversible Efficiency

The enthalpy of the overall chemical reaction in a fuel cell is the difference between the enthalpies of formation of the products and reactants. This enthalpy of formation represents the amount of heat energy produced from the complete combustion of the hydrogen fuel and will serve an important purpose when we define the reversible efficiency of a fuel cell. On a unit mole basis, the enthalpy of formation in a hydrogen fuel cell is given by

$$\Delta H_f = (h_f)_{H_2O} - (h_f)_{H_2} - (h_f)_{O_2} \tag{6}$$

The enthalpy of formation of elements such as oxygen and hydrogen is zero by definition while the enthalpy of formation of water can be calculated at different temperatures. Thus, Eq. (6) reduces to

$$\Delta H_f = (h_f)_{H_2O} \tag{7}$$

As per Eq. (3), if the product water is in liquid form, then we will be referring to the HHV of the enthalpy of formation. However, if the product water is in vapor form, then we are referring to the LHV of the enthalpy of formation. The difference between these two values is water's molar latent heat of vaporization.

However, since combustion oxidation does not occur within a fuel cell, the enthalpy of formation serves only as an indicator of the amount of energy input to a fuel cell. Neglecting work done for the change of pressure and/or volume, the maximum portion of the energy input to a fuel cell that could be converted into useful electric work is found from the Gibbs free energy of formation, which is given on a mole basis using:

$$\Delta G_f = \Delta H_f - T\Delta S_f$$

(8)

The entropy of formation for the reaction can be calculated in a similar way to the enthalpy of formation at different temperatures by the following equation:

$$\Delta S_f = (s_f)_{H_2O} - (s_f)_{H_2} - (s_f)_{O_2}$$

(9)

It is important to distinguish that ΔG_f is the maximum useful work associated with a chemical reaction while ΔH_f is the maximum heat associated with a chemical reaction. When all the ΔG_f is converted into useful electric work by moving electrons through an external circuit, the cell voltage is termed the reversible cell voltage. Finally, when considering Eq. (8), it is important to realize that the $T\Delta S_f$ term grows faster than the ΔH_f term with an increase in temperature. Thus, we expect ΔG_f to decrease in magnitude as temperature is increased.

In order to define a maximum efficiency concept similar to the Carnot efficiency in heat engines, we have to consider both the energy input to the fuel cell system and the maximum amount of that energy available to do external work. The former is the enthalpy of formation and the latter is the Gibbs free energy of formation, as per the above discussion. Thus, we can define the reversible maximum efficiency of a fuel cell as the ratio between these two variables, as follows:

$$\eta_{rev} = \frac{\Delta G_f}{\Delta H_f}$$

(10)

Fig. 22 compares the ideal theoretical efficiency of a fuel cell against that of a Carnot engine. We observe that at the lower and intermediate operating temperatures, the fuel cell ideal efficiency is significantly better. While at higher operating temperatures, both ideal efficiencies are close to each other. The practical implementation of a lower operating temperature is better dynamic response and quicker load-up times for applications. However, in reality, operation at higher operation temperatures results in higher cell voltage due to reduced voltage losses, as will be seen in later sections.

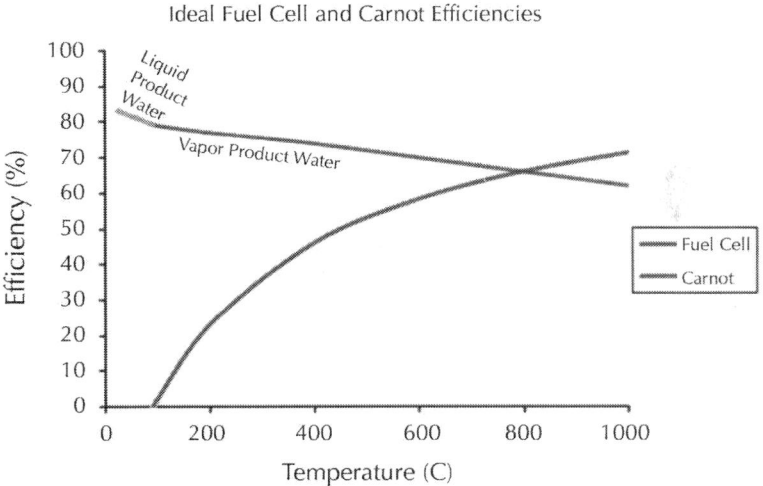

Figure 22: Ideal fuel cell (using LHV) and Carnot efficiencies (using exhaust temperature of 90 °C) as a function of temperature.

Reversible Voltage

Electrical work is defined as the product of charge and potential as

$$W_{ele} = qE \tag{11}$$

The total charge transferred with electrons in a fuel cell per every mole of H_2 is

$$q = n\, N_{avg}\, q_{el} \tag{12}$$

where n , N_{avg}, and q_{del} are the number of electrons per molecule of H_2 involved in the reaction, Avogadro's number, and the charge of an electron, respectively. Eq. (12) can be further simplified by noting that the product of N_{avg} and $_{qel}$ is the electric charge per mole of electrons, or in other words equal to Faraday's constant (96,485 C/electron-mole). Thus, the electrical work could be expressed as

$$W_{ele} = nFE$$

(13)

However, we saw that the Gibbs free energy of formation is equal to the electric work produced in a fuel cell when the system has no irreversibilities. Accordingly, W_{ele} is equal to $-\Delta G_f$ in a reversible system and we can express the reversible cell voltage in a fuel cell as

$$E_{rev} = -\frac{\Delta G_f}{nF}$$

(14)

The above value is the highest theoretically attainable voltage from an isothermal fuel cell and is commonly called the Nernst voltage. It is worth noting that if we replace the Gibbs free energy in Eq. (14) with enthalpy, we get what is known as the thermoneutral cell voltage, which corresponds to the complete conversion of all the energy content in the fuel to electric work (i.e., 100% thermal efficiency and no internal thermal energy generation). Substituting Eq. (8) into Eq. (14) yields:

$$E_{rev} = \frac{\Delta H_f - T\Delta S_f}{nF}$$

(15)

As we can see from Eq. (15) and from Fig. 23, increasing the temperature would result in a decrease in the theoretical potential of the fuel cell. However, mass transport and ionic conduction are faster at higher temperatures and this more than offsets the drop in the Nernst voltage, as will be seen in later sections. Using the definition of reversible cell voltage in Eq. (15), we can define the voltage efficiency of a fuel cell as

$$\eta_{vol} = \frac{E}{E_{rev}}$$

$$(16)$$

where E is the operating voltage. That is, the voltage efficiency is the ratio of the cell operating voltage to the Nernst voltage.

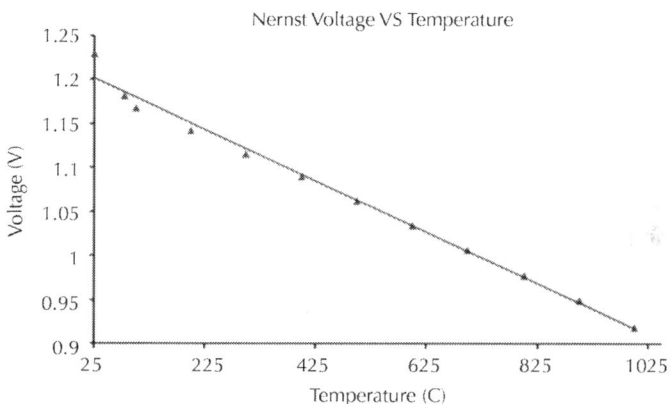

Figure 23: Nernst voltage as a function of temperature for a hydrogen/oxygen PEMFC at 1 atm reactants pressure.

So far, we have approximated the Gibbs free energy of formation as a function of temperature only, as per the discussions above. However, Gibbs free energy is a function of both temperature and pressure. This is evident when Gibbs free energy of formation for a hydrogen fuel cell is expressed according to the following more accurate thermodynamic equation:

$$\Delta G_f = \Delta G_f^0 - RT \ln \left(\frac{P_{H_2} P_{O_2}^{0.5}}{P_{H_2O}} \right)$$

$$(17)$$

where the 0 superscript indicates standard conditions and the P's are the gases partial pressures (proportional to the molar fractions in a mixture and assuming all species are in gaseous form). If we use the expression in Eq. (17) with the reversible voltage definition in Eq. (14), we get:

$$E_{rev} = E_{rev}^0 + \frac{RT}{nF} \ln\left(\frac{P_{H_2} P_{O_2}^{0.5}}{P_{H_2O}}\right)$$

(18)

where E_{rev}^0 is the reversible Nernst voltage at standard conditions. Thus, we observe from Eq. (18) that the partial pressures of the reactants and products play a significant role in changing the reversible Nernst cell voltage according to the concentration of the reactants and the products during the reaction. For instance, we can use Eq. (18) to find that the effect of changing the inlet pressure or the concentration of hydrogen (e.g., pure or with traces of COx), which will be equal to:

$$\Delta E_{rev} = \frac{RT}{nF} \ln \frac{(P_{H_2})_2}{(P_{H_2})_1}$$

(19)

Thus, when the reactants contain inert diluents, the diluents will cause a drop in the reversible cell voltage referred to as the Nernst loss. We can also use Eq. (18) to isolate the effect of changing the system's pressure from P_1 to P_2 on the reversible cell voltage, which will be equal to:

$$\Delta E_{rev} = \frac{RT}{2nF} \ln \frac{P_2}{P_1}$$

(20)

Finally, we can use Eq. (18) to identify the effect of changing from air to pure oxygen on the reversible cell voltage, which will be equal to:

$$\Delta E_{rev} = \frac{RT}{2nF} \ln \frac{1.0}{0.21}$$

(21)

Flow Rates

The amount of hydrogen and oxygen consumed in a fuel cell stack are a function of the current obtained from said stack. We can use Faraday's law to derive the relation between required flow rates of reactants for a specified current, where:

$$It = nzF \tag{22}$$

where I , t, n, z, and F are current in A, time in seconds, number of moles, number of electrons in the reaction, and Faraday's constant, respectively. Based on the reactions at the anode (Eq. (1)) and the cathode (Eq. (2)) for a hydrogen fuel cell and since z will be equal to 2 in this case, the molar flow rates of the reactants can be calculated as follows:

$$\dot{n}_{hydrogen} = \frac{I}{2F} \tag{23}$$

$$\dot{n}_{oxygen} = \frac{I}{4F} = \frac{\dot{n}_{hydrogen}}{2} \tag{24}$$

where $\dot{n}$ is the molar flow rate in mol s^{-1}. Taking into account stoichiometric ratios, number of cells per stack, and the generalized case where the fuel and oxidant are not pure; we get the following more practical equations for the required molar flow rates of fuel and oxidant given a certain current output:

$$\dot{n}_{fuel} = \frac{IS_{H_2} N_{cell}}{2Fr_{H_2}} \tag{25}$$

$$\dot{n}_{oxidant} = \frac{IS_{O_2} N_{cell}}{4Fr_{O_2}} \tag{26}$$

where N_{cell} is the number of unit cells, S is the stoichiometric ratio, and r is the volume/molar fraction. In order to determine the molar flow rate of the water content in the fuel exhaust $((\dot{n}_{H_2O})_{fuel.out})$, we observe that it is going to be equal to the water content in the fuel inlet $((\dot{n}_{H_2O})_{fuel.in})$, plus the water transported from the cathode to the anode as a result of back diffusion $((\dot{n}_{H_2O})_{BD})$, less the water transported from the anode to the cathode as a result of electroosmotic drag $((\dot{n}_{H_2O})_{ED})$. Accordingly

$$(\dot{n}_{H_2O})_{fuel,\ out} = (\dot{n}_{H_2O})_{fuel,\ in} + (\dot{n}_{H_2O})_{BD} - (\dot{n}_{H_2O})_{ED} \tag{27}$$

Similarly, the molar flow rate of the water content in the oxidant exhaust $((\dot{n}_{H_2O})_{oxidant.out})$ is going to be equal to the water content in the oxidant inlet $((\dot{n}_{H_2O})_{oxidant.in})$, plus the water generated at the cathode $((\dot{n}_{H_2O})_{gen})$, plus the water transported from the anode to the cathode as a result of electroosmotic drag $((\dot{n}_{H_2O})_{ED})$, less the water transported from the cathode to the anode as a result of back diffusion $((\dot{n}_{H_2O})_{BD})$. Accordingly

$$(\dot{n}_{H_2O})_{oxidant,\ out} = (\dot{n}_{H_2O})_{oxidant,\ in} + (\dot{n}_{H_2O})_{gen} + (\dot{n}_{H_2O})_{ED} - (\dot{n}_{H_2O})_{BD} \tag{28}$$

Usually, we express back diffusion as a fraction of electroosmotic drag across a membrane. This is because back diffusion depends (in addition to other factors) on the water concentration gradient across the membrane, which is not uniform and difficult to model. While on the other hand, electroosmotic drag depends only on the current drawn. Accordingly, we can use the following to find both terms:

$$(\dot{n}_{H_2O})_{ED} = \xi\frac{IN_{cell}}{F} \tag{29}$$

$$(\dot{n}_{H_2O})_{BD} = \beta\xi\frac{IN_{cell}}{F} \tag{30}$$

where ξ is number of water molecules per proton and β is the fraction used to express back diffusion in terms of electroosmotic drag.

Fuel Cell Polarization

As mentioned earlier, the reversible cell voltage is the voltage that can be obtained if the Gibbs free energy could be converted directly into electrical work without any losses. However, in reality, there are several irreversibilities within a fuel cell that cause the actual cell voltage to be less than the reversible cell voltage. These irreversibilities cause the actual voltage to decline as current density increases. Thus, it is useful to plot cell voltage against current density as a merit of characterization for a certain fuel cell. And even at the open-circuit voltage state where no load exists, the actual voltage is still less than the reversible voltage. These irreversibilities are known as cell polarizations and could be divided into four main polarization sources; namely, crossover, activation, ohmic, and concentration losses, as depicted in Fig. 24. These polarization sources are active throughout the entire polarization curve. However, they become dominant at certain segments of the polarization curve, as will be shown in the following discussions. The polarization curve shown in Fig. 24 is one of the most important merits of evaluation in fuel cell science and when the four main polarizations are deducted from the reversible voltage we get what is known as the polarization equation:

$$E = E_{rev} - E_{a,\,a} - E_{a,\,c} - E_o - E_{c,\,a} - E_{c,c} \tag{31}$$

where $E_{a.a}$ and $E_{a.c}$ are the activation and crossover losses at the anode and cathode, E_o are the ohmic losses, and $E_{c.a}$ and $E_{c.c}$ are the concentration losses at the anode and cathode. All the terms in Eq. (31) need to be positive.

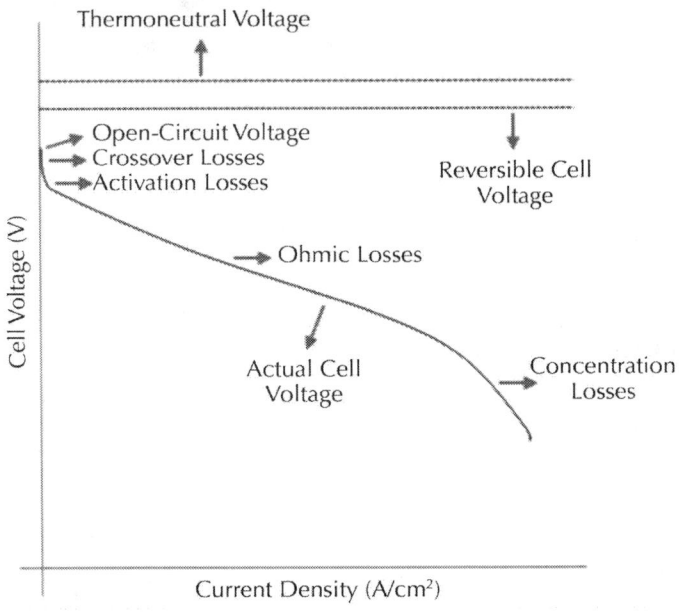

Figure 24: Typical polarization curve with voltage losses.

Activation polarization is the main cause of voltage drop at low current densities and is caused by sluggish oxidation and/or reduction kinetics at the electrodes surface. Initiating the electrochemical reactions requires energy that is reflected in the activation voltage drop. The activations losses at the anode and cathode could be isolated and expressed using Tafel's equation:

$$E_{a,\,a} = A_a \ln\left(\frac{i}{i_{0,\,a}}\right)$$

(32)

$$E_{a,\,c} = A_c \ln\left(\frac{i}{i_{0,\,c}}\right)$$

(33)

where i is the current density, i_0 is the exchange current density, and A is given from:

$$A = \frac{RT}{n\alpha F}$$

(34)

where α is a constant known as the charge transfer coefficient which depends on the electrode›s material, microstructure, and reaction mechanism. The exchange current density is defined as the rate at which the simultaneous oxidation and reduction reactions occur under equilibrium conditions when the net current is zero. Thus, it is a measure of the electrode›s activity and the higher its value, the easier it is for a charge to move from/to the electrode to/from the electrolyte and the greater the current density. The exchange current density is the determining factor in activation losses. Its value is best given using the following equation [114]:

$$i_o = i_0^{ref} \varepsilon_c P_r^\gamma \exp\left(-\frac{E_c}{RT}(1 - T_r) \right)$$

(35)

where i^{ref}_0 is the exchange current density at arbitrary reference conditions, ε_c is the electrode reference (typically between 180 and 500), P_r is the ratio between the reactant partial pressure and the reactant reference pressure, γ is the pressure coefficient (typically between 0.5 and 1.0), $_{Ec}$ is the activation energy (equal to 66 kJ mol^{-1} for oxygen reduction on platinum), and $_{Tr}$ is the ration between the temperature and the reference temperature.

We always desire to maximize the value of the exchange current density in order to minimize the activation losses, as evident from Eqs. (32) and (33). As evident from Eq. (35), we can increase the value of the exchange current density by choosing an active electrode catalyst, increasing the operation temperature, increasing the operation pressure, increasing the roughness of the surface area of the electrodes to increase the active reaction sites, increasing the catalyst loading, and increasing the reactants concentration to increase the active spots on the electrodes surface area. Other factors that increase the activation losses due to catalyst degradation include the presence of catalyst contaminants in the reactants and prolonged loading cycles. It is important to notice that different activation losses occur at each electrode, hence, the two different terms in Eq. (31). Nevertheless, the current that passes through the two electrodes is the same. In hydrogen fuel cells, the reduction activation losses dominate. This is because the exchange current density for the reduction reaction is much less than that for the oxidation reaction. Accordingly, we are mostly concerned with increasing the exchange current density for the reduction reaction. It is interesting to notice that even though we witnessed previously that at higher temperatures the reversible cell voltage is lower; this is not the case for the actual cell voltage. The fact that activation losses decrease as temperature increases (as a result of the increased exchange current density, as per Eqs.(32), (33) and (35)) causes the actual cell voltage to actually increase with increased temperature. And even though the value of A in Eq. (34) will increase with increased temperatures, the increase in the exchange current density more than offsets the increase in A. Accordingly, the end result is usually less activation voltage losses as a result of increased temperatures. Thus, the actual cell voltage curve is shifted up while the reversible cell voltage is shifted down as a result of increased temperatures and the gap between them is lessened.

Crossover polarization is the main cause of voltage loss at open-circuit conditions and is due to two reasons. The first is the direct hydrogen fuel diffusion from the anode to the cathode through the electrolyte without the anodic reaction taking place, even though the membrane is practically impermeable to the hydrogen fuel. The second is the internal passing of electrons through the electrolyte rather than through the external circuit, even though the membrane is practically impermeable to electrons. The effect of these two phenomena is the same on the cell voltage (i.e., a voltage drop due to wasted hydrogen fuel and/or electrons). Crossover polarization is usually noticeable when the operation temperature is low and we are at or near the open-circuit conditions. This is because at open-circuit the only hydrogen fuel consumption occurring is due to crossover polarization while at closed-circuit the hydrogen consumed to generate the useful external current is much greater than that consumed (or wasted) as a result of the internal current. Moreover, the hydrogen concentration gradient across the membrane decreases with higher current densities due to the higher rates of hydrogen consumption at the anode. Thus, the driving force for the hydrogen molecules to diffuse through the membrane (i.e., the hydrogen concentration gradient) is weak at high current densities and strongest when no external current is drawn (i.e., at open-circuit conditions). As a result, crossover losses could be isolated and measured by measuring the small reactants consumption at open-circuit conditions when no external current is running.

So if we use Eq. (23) and divide by cell active area, we can express the internal current density at open-circuit conditions due to crossover polarization in a hydrogen fuel cell as

$$i_{loss} = \frac{2F\dot{n}_{hydrogen}}{A}$$

(36)

Accordingly, the lost internal current density due to crossover polarization could be incorporated into Eqs.(32) and (33) to yield:

$$E_{a,\,a} = A_a \ln \left(\frac{i_{loss} + i}{i_{0,\,a}} \right)$$

(37)

$$E_{a,\,c} = A_c \ln \left(\frac{i_{loss} + i}{i_{0,\,c}} \right)$$

(38)

This explains why the actual voltage is always less than the reversible voltage, as per Eq. (31). That is, the terms in Eqs. (37) and (38) will not be equal to zero at open-circuit conditions due to the lost internal currents. As a result, the open-circuit voltage will always be less than the reversible Nernst voltage, as depicted in Fig. 24. Nevertheless, it is possible to minimize crossover losses by optimizing the membrane's permeability and thickness.

The ionic and electric resistance of the stack's components to the flow of charge results in ohmic polarization. The electrolyte, catalyst layer, GDL, flow field plates, current collectors, interfacial contacts between the components, and the terminal connections all contribute to these ohmic voltage losses. The electric resistivity is due to the resistivity of the electrically-conductive cell components to the electrons flow while the ionic resistivity is due to the resistivity of the membrane to the ions flow. Most of the electric resistivity occurs due to the lack of proper contact between the GDL, bipolar plates, cooling plates, and other interconnects. However, usually, the ionic resistivity dominates ohmic voltage losses. This is because the number of charge carriers through an ionic conductor is much less than in an electronic conductor. In an electronic conductor, the valence electrons of the atoms become detached and can move freely, whereas in ionic conductors, the ions move through the vacancies in the crystallographic lattice. Thus, the electronic resistance is usually negligible in comparison to the ionic and contact resistances.

We can express the ohmic voltage losses due to ionic, contact, and electronic resistances according to Ohm's law as

$$E_0 = i(R_{ele} + R_{ion} + R_{CR}) \tag{39}$$

where R_{ele}, R_{ion}, and R_{CR} are the area-specific electronic, ionic, and contact resistances in $\Omega \, cm^2$. Ohmic losses are dominant at the middle of the polarization curve and affect all types of fuel cells. Thus, in order to minimize the ohmic losses, it is important to design the stack from materials with high conductivities (i.e., low resistivities), components with minimum thicknesses, and interconnects with minimum contact resistances through the optimization of the stack's compression pressure. This is particularly important for the electrolyte due to its dominant ionic resistivity. This could be achieved by designing a chemically and mechanically stable electrolyte with the highest possible conductivity and the smallest possible thickness since the resistivity of the electrolyte is proportional to the ratio of its thickness over conductivity. Also, the electrolyte material and water content play a significant role in determining its resistivity and need to be carefully considered.

Concentration polarization is dominant at high current densities and occurs when the electrode reactions are hindered by reduced reactants availability (i.e., concentration) at reaction sites. This concentration reduction (which translates to a partial pressure reduction) could be due to limited hydrogen fuel supply, limited diffusion rate of the fuel and oxidant from flow field channels to the catalyst layer, poor air circulation at the cathode which leads to nitrogen (or any other nonparticipating inert gases for that matter) build-up, water accumulation and flooding at the cathode and anode (especially for PEMFCs), or impurities adsorption on electrode reaction areas. However, the cathode's concentration polarization usually dominates since water accumulation usually occurs at the cathode, nitrogen build-up also usually occurs at the cathode, and the diffusion rate of oxygen is much lower than that of hydrogen. It is possible to minimize concentration voltage losses by proper water management, removal of impurities, optimizing the stoichiometric ratio, and optimizing the thickness and porosity of the GDL.

In order to describe the concentration voltage losses, we note that the maximum current density the fuel cell can produce occurs when the rate of reactant (i.e., the fuel or the oxidant) consumption is equal to the rate of reactant supply. Thus, at this maximum current density the concentration of the reactant (i.e., its partial pressure) at the surface of the catalyst would reach zero. Similarly, the maximum concentration of the reactant (i.e., its maximum partial pressure) occurs when the current density drawn is zero. Assuming we have a linear relationship between the partial pressure of the reactant and current density generated, we come up with the following simple linear equation that relates the two variables (applicable to fuel and oxidant):

$$P = -\frac{P_{max}}{i_{max}}i + P_{max}$$

$$(40)$$

where P_{max} is the maximum partial pressure corresponding to the maximum concentration, i_{max} is the maximum current density, P is any pressure between zero and P_{max}, and i is any current density between zero and i_{max}. By rearranging we obtain

$$\frac{P}{P_{max}} = 1 - \frac{i}{i_{max}}$$

(41)

Recall the relations we established in Eqs. (18), (19) and (21) based on the Nernst voltage concept to describe how the variation of a reactant partial pressure affects the voltage. Based on these relations and by replacing the $_{P_2}/_{P_1}$ terms with the P/P_{max} expression we have in Eq. (41), we establish the concentration voltage losses at the anode and cathode with hydrogen and oxygen flows as follows:

$$E_{c,\,a} = -\frac{RT}{2F} \ln \left(1 - \frac{i}{i_{max,\,a}} \right)$$

(42)

$$E_{c,\,c} = -\frac{RT}{4F} \ln \left(1 - \frac{i}{i_{max,\,c}} \right)$$

(43)

Notice the addition of the negative sign so that the outcome is a positive voltage loss value. Even though Eqs. (42) and (43) are a fair fit, they lack the gradual smoothness of actual experimental results. This is because these two equations do not take into account the effect of water and non-reacting inert gases (especially nitrogen) accumulation which leads to non-uniform current density distribution over the electrode surface area. Hence, we usually use an empirical constant (B) in Eqs. (42) and (43) to give us more accurate results. Accordingly

$$E_{c,\,a} = -B_a \ln \left(1 - \frac{i}{i_{max,\,a}} \right)$$

(44)

$$E_{c,\,c} = -B_c \ln \left(1 - \frac{i}{i_{max,\,c}} \right)$$

(45)

It is worth noting that as seen from Fig. 24, the constant difference between the thermoneutral voltage and the reversible cell voltage is due to the $T\Delta S_f$ term previously discussed in Eq. (8). This constant difference represents the minimum amount of fuel input energy that must be converted into thermal energy under ideal fuel cell conditions. This is analogous to the Carnot efficiency concept in heat engines that represents the minimum amount of input energy that needs to be converted into thermal energy between a source and a reservoir at known temperatures. The difference between the thermoneutral voltage and the actual cell voltage, as depicted in Fig. 24, represents the actual amount of heat generation within the fuel cell. When this difference is multiplied by the current density, we get what is known as the heat generation density rate curve.

The polarization curve and equation represent a zero-dimensional steady-state model for a hydrogen fuel cell under the assumption that only a single gaseous phase is present. This is one of the simplest and most common tools for the evaluation of fuel cell performance. Nevertheless, more involved multi-dimensional and multi-phase models exist where numerical iterations and software packages are used. Based on the discussions in the previous section, we will generate the polarization curve of a typical hydrogen-air PEMFC and breakdown the curve into its crossover, activation, ohmic, and concentration loss curves. We will also generate the power density curve and the density rate of heat generation curve for the same fuel cell.

If we use the expressions found in Eqs. (18), (37), (38), (39), (44), and (45) for the variables in Eq. (31), the result is the following polarization equation:

$$E = \left[E_{rev}^0 + \frac{RT}{nF} \ln \left(\frac{P_{H_2} P_{O_2}^{0.5}}{P_{H_2O}} \right) \right] - \left[A_a \ln \left(\frac{i_{loss} + i}{i_{0,a}} \right) \right] - \left[A_c \ln \left(\frac{i_{loss} + i}{i_{0,c}} \right) \right]$$

$$- [i(R_{ele} + R_{ion} + R_{CR})] - \left[-B_a \ln \left(1 - \frac{i}{i_{max,a}} \right) \right]$$

$$- \left[-B_c \ln \left(1 - \frac{i}{i_{max,c}} \right) \right]$$

$$(46)$$

The parameters in this equation are listed in Table 15 for a typical hydrogen-air PEMFC. The values for A_a and A_c were calculated using Eq. (34) and the values in Table 15. It is important to realize that all the voltage loss terms within the square brackets in this equation are positive. Fig. 25 shows the polarization curve of the aforementioned

PEMFC with voltage losses breakdown. In accordance with the previous discussions, it is clear that the activation losses dominate at low current densities with crossover losses responsible for the losses at zero current density. The ohmic losses linearly increase with increased current densities and dominate the intermediate range with the activation losses. While the concentration losses are very low until we reach the high current densities region where they dominate and are responsible for bringing the cell voltage to zero as a result of the current density reaching the maximum current density. Fig. 26 shows three out of four of the most important fuel cell performance evaluation curves (the fourth being the efficiency curve). The figure shows the opposed relation between the polarization and power density curves on the one hand and the density rate of heat generation curve on the other. The input fuel energy that is not being converted into useful electric energy is wasted as internal stack thermal energy. The power density curve shows a wide optimum range of current densities where power density is at its near-peak. This is an important observation for the stack designer and user.

Table 15: Parameter values for the polarization, power, and heat generation curves

Parameter	Value	Unit	Parameter description
ΔH_f	285,250	J mol^{-1}	Enthalpy of formation
E^0_{rev}	1.18	V	Reversible Nernst voltage at standard conditions
R	8.3145	J mol^{-1} K^{-1}	Gas constant
T	353	K	Operation temperature
n	2	–	Number of electrons involved
F	96,485	C mol^{-1}	Faraday's constant
P_{H_2}	100	kPa	Hydrogen partial pressure
P_{O_2}	21	kPa	Oxygen partial pressure

P_{H_2O}	45	kPa	Water partial pressure
α_a	0.5	–	Anode charge transfer coefficient
α_c	0.3	–	Cathode charge transfer coefficient
A_a	0.03042	V	Anode activation constant
A_c	0.05070	V	Cathode activation constant
$i_{0.a}$	0.15	A cm^{-2}	Anode exchange current density
$i_{0.c}$	1.5×10^{-4}	A cm^{-2}	Cathode exchange current density
i_{loss}	0.008	A cm^{-2}	Lost internal current density
R_{ele}	0.0	$\Omega\,cm^2$	Area-specific electronic resistance
R_{ion}	0.10	$\Omega\,cm^2$	Area-specific ionic resistance
R_{CR}	0.030	$\Omega\,cm^2$	Area-specific contact resistance
B_a	0.045	V	Anode empirical constant
B_c	0.045	V	Cathode empirical constant
$i_{max.a}$	15	A cm^{-2}	Anode maximum current density
$i_{max.c}$	2.5	A cm^{-2}	Cathode maximum current density

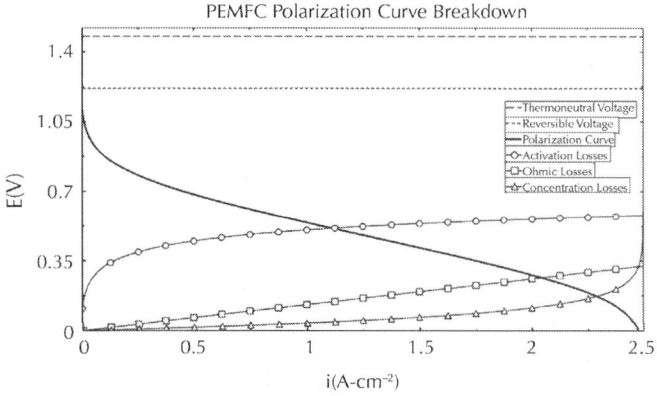

Figure 25: PEMFC polarization curve breakdown.

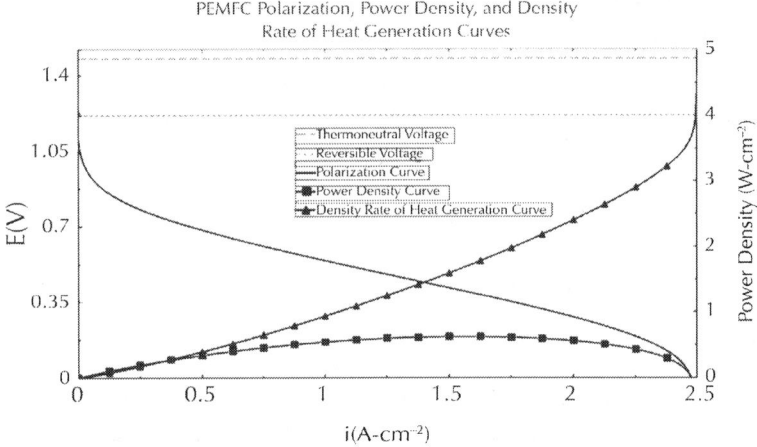

Figure 26: PEMFC polarization, power density, and density rate of heat generation curves.

The equations of the fuel cell zero-dimensional polarization model presented so far assume static variables (such as reactant flow rates); while in practice many of the variables are in flux and change with time. Hence, for practical fuel cell design, control, stability analysis, evaluation, and optimization; dynamic modeling is required. This could be realized by dividing the terms in Eq. (31) into two main components. The first being the steady components (E_{st}) and the second being the transient components (E_{tr}). The steady components include the reversible voltage and the ohmic voltage losses since they are independent of transient terms. The transient components include the more involved activation and concentration voltage losses since they depend on transient terms. Accordingly:

$$E = [E_{st}] - [E_{tr}] = [E_{rev} - E_o] - [E_{a,\,a} + E_{a,\,c} + E_{c,\,a} + E_{c,\,c}] \tag{47}$$

Finding expressions for the steady and transient components in Eq. (47) could be achieved using different models available in the literature such as the fuel cell unified mathematical model described in Ref. [115].

Fuel Cell System Efficiency

The overall fuel cell system efficiency consists of a series of efficiencies. First of which is the reversible efficiency that was discussed in Section 7.1 and is given by Eq. (10). The second is the voltage efficiency that was discussed in Section 7.2 and is given by Eq. (16). Additionally, the fuel utilization efficiency (u_{fuel}) is the fraction of the fuel consumed within a fuel cell, the power conditioning efficiency ($_{pc}$) is the efficiency of the device used to condition the output power, the onboard reformer efficiency (u_{ref}) is the fraction of the raw fuel transformed into fuel cell usable fuel, and the parasitic power efficiency takes into account the amount of fuel cell power used to operate the BoP subsystems, which is given by the following semi-empirical equation:

$$\eta_P = 1 - a - \frac{b}{Ei}$$

(48)

where a and b are empirical constants. When all the previously-mentioned efficiencies are combined, we get the overall fuel cell system efficiency as follows after simplification:

$$\eta_{tot} = \frac{nFE}{\Delta H_f}(u_{fuel}u_{ref}\eta_{pc})\left(1 - a - \frac{b}{Ei}\right)$$

(49)

By substituting the cell actual voltage in Eq. (46), using the hydrogen/air PEMFC from the previous section, into Eq. (49), we generate the total system efficiency curve in Fig. 27. Table 16 lists the parameters used in Eq. (49). We observe from the figure that for the used parameter values, the efficiency is highest around 0.5 current density. The efficiency is also very low at near-zero current densities and linearly decreases between 0.5 and 2 current densities then exponentially drops between 2 and 2.5 current densities. This implies that it is possible to optimize the design of a fuel cell by creating optimum ranges for the design parameters so as to remain within the optimum efficiency range.

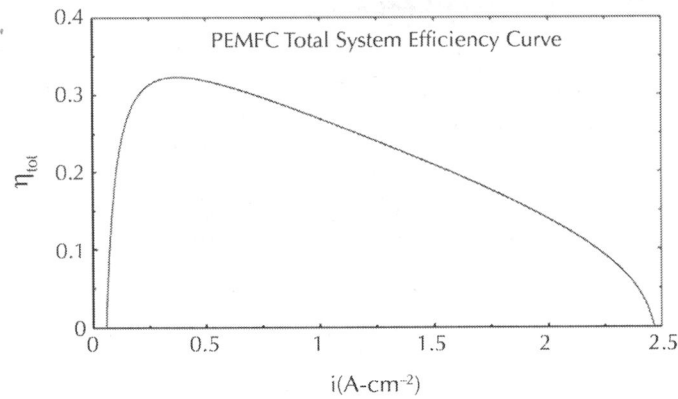

Figure 27: PEMFC total system efficiency curve.

Table 16: Parameter values for the total system efficiency curve

Parameter	Value	Unit
u_{fuel}	0.9	–
u_{ref}	1	–
η_{pc}	0.95	–
a	0.0499	W m^{-1}
b	0.05	–

FUEL CELL SYSTEMS EVALUATION FACTORS

The evaluation and comparison of different fuel cell stacks for a certain given application is dependent on several factors. Many of these factors are case-dependent; however, it is possible to group them into the following:

- Physical factors include total stack size and weight, cell active area, number of cells, number of stacks, and total BoP subsystems size and weight. These are important factors when the size and/or weight of a fuel cell system are design constraints that need to be met. This is most commonly experienced for transportation and portable applications when the weight and/or size of the fuel cell system are critical constrains, as we witnessed in Section 4.

- Performance factors include the polarization, power density, and system efficiency curves. As evident from the previous section, it is possible to optimize the stack design by restricting the design parameters to ranges that fall into the peak power density and/or system efficiency regions. Additionally, the power density curve provides an indication of the flexibility of the output power the stack can produce. While the polarization curve, in combination with the power density curve, is usually used to determine optimum operation points in terms of voltage, current, and power.

- Running costs are a function of stack fuel consumption, system parasitic loads, thermal management system requirements, and efficiency of the power conditioning equipment; in addition to other case-dependent factors.

- Durability is also a critical factor in choosing a fuel cell stack for a certain application. This is most manifest for stationary power generation applications where the fuel cell system is expected to function for a reasonable amount of time with minimum maintenance requirements, as previously discussed in Section 4.

PROSPECTS AND CONCLUDING REMARKS

This study is a brief summary of the state of up-to-date fuel cell technologies, with a stress on the underlying but key link between application possibilities, features and characteristics, and principals of operation. An overlook over the history of fuel cells technology, competing power generation technologies, and fuel cell types have been provided. The characterizing advantages and disadvantages of fuel cells have been briefly described leading to a review of the most recent fuel cell pilot deployment and demonstration initiatives along with market status and prospects in the portable, stationary, and transportation sectors. After that, current progress and future targets for the fuel cell community have been identified in order to accelerate the market penetration rate of fuel cells. Finally, the manuscript covered the design levels, thermodynamic and electrochemical fundamentals, and system evaluation factors to offer the reader an insight into the design and operation details of fuel cell systems. Throughout the text, the authors explored global paths to the future with the aim of making fuel cells an economically-competitive player in the energy market.

A parallel approach is recommended in order to penetrate the market with a focus on DMFCs in the consumer electronics market, PEMFCs in the personal passenger vehicles market, and high-temperature fuel cells for the distributed CHP and trigeneration market. Nevertheless, the economic feasibility and reliability of fuel cells as alternatives to conventional power generation solutions are compromised by their high cost and low durability. And despite several successful pilot projects, pre-commercial demonstration initiatives, and niche market penetrations in the US, Japan, and EU; research, development, and deployment efforts should mainly target resolving the cost and durability issues before fuel cells can achieve a reasonable penetration rate into the portable, stationary, and transportation markets. Based on the discussions and findings of this study, reducing the high cost of fuel cells could be accomplished by reducing PGM loading; developing MEA fabrication processes suitable for mass production; developing durable and economic bipolar plate materials; developing economic BoP components; and simplifying fuel processing and reformation units. At the same time, increasing the durability of fuel cells in order to stand as reliable alternatives to current power generation technologies could be accomplished by improving the MEA's contamination tolerance; developing MEA material capable of handling wide ranges of operation conditions and load dynamics; fundamentally understanding and comprehensively modeling water transport phenomena so as to develop advanced materials and optimized water management techniques; developing sealing, bipolar plate, and MEA materials and designs customized for high temperature operation; and developing mitigation strategies to handle the corrosion of platinum and its carbon support. Other issues in need of dire attention include dynamic transient operation models validated with experimental data, auxiliary components specifically optimized for fuel cell integration, standardized testing and characterizing techniques, and stack-level optimization and testing. At the end, the resolution of the cost and durability issues is likely to be directly reflected as an accelerated institution of the hydrogen infrastructure and a more accepting public, political, and industrial reception.

Indeed, fuel cells offer the highly-desirable combination of high efficiency and environmentally-benign operation that most other energy conversion devices lack; nonetheless, the development and commercialization of fuel cells have been a long and irresolute process. This is highly due to the fact that the emergence of heat engines, batteries, and similar devices has often overshadowed fuel cells for the simple fact that we have often been engrossed by the cost, efficiency, and reliability of energy generation and conversion technologies at the expense of the environmental aspect. However, as a result of the growing environmental concerns that world leaders are finally paying attention to, a revived interest in hydrogen and fuel cell technologies started during the 1990s. And as a result, more technological advances and successful commercial endeavors took place during the past two decades compared to the period between 1839, when Grove and Schönbein demonstrated their fuel cell concepts, and the 1990s. This only calls for more attention from the academic community and more support from the public and private sectors to solve the current technological challenges facing hydrogen systems in general and fuel cells in particular in order to soon realize the full potential and benefits of a hydrogen economy.

REFERENCES

1.   Polymer electrolyte membrane fuel cells. Helsinki University of Technology. ⟨http://tfy.tkk.fi/aes/AES/projects/renew/fuelcell/pem_index.html⟩.

2.   Space applications of hydrogen and fuel cells. National Aeronautics and Space Administration. ⟨http://www.nasa.gov/topics/technology/hydrogen/ hydrogen_2009.html⟩.

3.   von Spakovsky MR, Olsommer B. Fuel cell systems and system modeling and analysis perspectives for fuel cell development. Energy Convers Manage 2002;43:1249–57.

4.   Gray EM, Webb CJ, Andrews J, Shabani B, Tsai PJ, Chan SLI. Hydrogen storage for off-grid power supply. Int J Hydrog Energy 2011;36:654–63.

5.   Srinivasan S. Fuel cells: from fundamentals to applications. New York: Springer; 2006.

6.   Hydrogen fuel cell engines and related technologies course manual. US Department of Energy. ⟨http://www1.eere.energy.gov/hydrogenandfuelcells/   tech_validation/pdfs/fcm04r0.pdf⟩.

7.   Sheffield JW, Sheffield Ç, editors. Assessment of hydrogen energy for sustainable development. Netherlands: Springer; 2007.

8.   Gaines L.L., Elgowainy A., Wang M.Q., Full fuel-cycle comparison of forklift propulsion systems. Argonne National Laboratory. ANL/ESD/08-3; 2008.

9.   The Department of Energy hydrogen and fuel cells program plan: an integrated strategic plan for the research, development, and demonstration of hydrogen and fuel cell technologies. US Department of Energy; 2011.

10.   El-Sharkh MY, Rahman A, Alam MS, Byrne PC, Sakla AA, Thomas T. A dynamic model for a stand-alone PEM fuel cell power plant for residential applications. J Power Sour 2004;138:199–204.

11. Zhu Y, Tomsovic K. Development of models for analyzing the load—following performance of microturbines and fuel cells. Electr Power Syst Res 2002;62:1–11.

12. Matheny MS, Erickson PA, Niezrecki C, Roan VP. Interior and exterior noise emitted by a fuel cell transit bus. J Sound Vib 2002;251:937–43.

13. Karlström M. Local environmental benefits of fuel cell buses—a case study. J Clean Prod 2005;13:679–85.

14. Grimes CA, Varghese OK, S. Ranjan. Light, water, hydrogen: the solar generation of hydrogen by water photoelectrolysis. New York: Springer; 2008.

15. Jiao K, Li X. Water transport in polymer electrolyte membrane fuel cells. Prog Energy Combust Sci 2011;37:221–91.

16. Ous T, Arcoumanis C. Degradation aspects of water formation and transport in proton exchange membrane fuel cell: a review. J Power Sour 2013;240: 558–82.

17. Tsushima S, Hirai S. In situ diagnostics for water transport in proton exchange membrane fuel cells. Prog Energy Combust Sci 2011;37:204–20.

18. Varkaraki E, Lymberopoulos N, Zachariou A. Hydrogen based emergency back-up system for telecommunication applications. J Power Sources 2003;118:14–22.

19. Elgowainy A, Gaines L, Wang M. Fuel-cycle analysis of early market applications of fuel cells: forklift propulsion systems and distributed power generation. Int J Hydrog Energy 2009;34:3557–70.

20. Renquist JV, Dickman B, Bradley TH. Economic comparison of fuel cell powered forklifts to battery powered forklifts. Int J Hydrog Energy 2012;37: 12054–9.

21. Airport Cooperative Research Program, CDM Federal Programs Corporation, KB Environmental Sciences, Ricondo & Associates. Airport ground support equipment (GSE): emission reduction strategies, inventory, and tutorial. Transportation Research Board of the National Academies. Report 78; 2012.

22. Tollefson J. US Congress revives hydrogen vehicle research. Nature 2009;460:442–3.

23. Carter D, Ryan M, Wing J. The fuel cell industry review 2012. Fuel Cell Today 2012.

24. Cowey K, Green KJ, Mepsted GO, Reeve R. Portable and military fuel cells. Curr Opin Solid State Mater Sci 2004;8:367–71.

25. Patil AS, Dubois TG, Sifer N, Bostic E, Gardner K, Quah M, et al. Portable fuel cell systems for America0 s army: technology transition to the field. J Power Sources 2004;136:220–5.

26. Horizon Fuel Cell Technologies. 〈http://www.horizonfuelcell.com〉.

27. Heliocentris. 〈http://www.heliocentris.com〉.

28. Wang Y, Chen KS, Mishler J, Cho SC, Adroher XC. A review of polymer electrolyte membrane fuel cells: technology, applications, and needs on fundamental research. Appl Energy 2011;88:981–1007.

29. Breakthrough Technologies Institute. 2011 Fuel cell technologies market report. US Department of Energy. DOE/EE-0755; 2012.

30. Liming H. Financing rural renewable energy: a comparison between China and India. Renew Sustain Energy Rev 2009;13:1096–103.

31. Abdullah MO, Yung VC, Anyi M, Othman AK, Ab. Hamid KB, Tarawe J. Review and comparison study of hybrid diesel/solar/hydro/fuel cell energy schemes for a rural ICT Telecenter. Energy 2010;35:639–46.

32. Bauen A, Hart D, Chase A. Fuel cells for distributed generation in developing countries—an analysis. Int J Hydrog Energy 2003;28:695–701.

33. Contreras A, Posso F, Guervos E. Modelling and simulation of the utilization of a PEM fuel cell in the rural sector of Venezuela. Appl Energy 2010;87:1376–85.

34. Munuswamy S, Nakamura K, Katta A. Comparing the cost of electricity sourced from a fuel cell-based renewable energy system and the national grid to electrify a rural health centre in India: a case study. Renew Energy 2011;36:2978–83.

35. Santarelli M. Design and analysis of stand-alone hydrogen energy systems with different renewable sources. Int J Hydrog Energy 2004;29:1571–86.

36. Khan MJ, Iqbal MT. Pre-feasibility study of stand-alone hybrid energy systems for applications in Newfoundland. Renew Energy 2005;30:835–54.

37. Wallmark C, Alvfors P. Technical design and economic evaluation of a standalone PEFC system for buildings in Sweden. J Power Sources 2003;118:358–66.

38. Agbossou K, Chahine R, Hamelin J, Laurencelle F, Anouar A, St-Arnaud JM, et al. Renewable energy systems based on hydrogen for remote applications. J Power Sources 2001;96:168–72.

39. Shabani B, Andrews J. An experimental investigation of a PEM fuel cell to supply both heat and power in a solar-hydrogen RAPS system. Int J Hydrog Energy 2011;36:5442–52.

40. Perrin M, Lemaire-Potteau E. Applications: stationary: remote area power supply: batteries and fuel cells. In: Garche J, editor. Encyclopedia of electrochemical power sources. Amsterdam: Elsevier; 2009.

41. Zoulias EI, Lymberopoulos N. Techno-economic analysis of the integration of hydrogen energy technologies in renewable energy-based stand-alone power systems. Renew Energy 2007;32:680–96.

42. Briguglio N, Ferraro M, Brunaccini G, Antonucci V. Evaluation of a low temperature fuel cell system for residential CHP. Int J Hydrog Energy 2011;36:8023–9.

43. Gigliucci G, Petruzzi L, Cerelli E, Garzisi A, La Mendola A. Demonstration of a residential CHP system based on PEM fuel cells. J Power Sources 2004;131: 62–8.

44. Kazempoor P, Dorer V, Weber A. Modelling and evaluation of building integrated SOFC systems. Int J Hydrog Energy 2011;36:13241–9.

45. Burer M, Tanaka K, Favrat D, Yamada K. Multi-criteria optimization of a district cogeneration plant integrating a solid oxide fuel cell–gas turbine combined cycle, heat pumps and chillers. Energy 2003;28:497–518.

46. Margalef P, Samuelsen S. Integration of a molten carbonate fuel cell with a direct exhaust absorption chiller. J Power Sources 2010;195:5674–85.

47. Chan CW, Ling-Chin J, Roskilly AP. A review of chemical heat pumps, thermodynamic cycles and thermal energy storage technologies for low grade heat utilization. Appl Therm Eng 2013;50:1257–73.

48. Bendaikha W, Larbi S, Bouziane M. Feasibility study of hybrid fuel cell and geothermal heat pump used for air conditioning in Algeria. Int J Hydrog Energy 2011;36:4253–61.

49. Ma S, Wang J, Yan Z, Dai Y, Lu B. Thermodynamic analysis of a new combined cooling, heat and power system driven by solid oxide fuel cell based on ammonia–water mixture. J Power Sources 2011;196:8463–71.

50. Wang M, Elgowainy A, Han J. Life-cycle analysis of criteria pollutant emissions from stationary fuel cell systems. US Department of Energy Hydrogen and Fuel Cells Program. AN012;2010.

51. Alcaide F, Cabot PL, Brillas E. Fuel cells for chemicals and energy cogeneration. J Power Sources 2006;153:47–60.

52. Tanino T, Nara Y, Tsujiguchi T, Ohshima T. Coproduction of acetic acid and electricity by application of microbial fuel cell technology to vinegar fermentation. J Biosci Bioeng 2013;116:219–23.

53. Plunkett JW. Plunkett0 s automobile industry almanac 2011. Houston, TX: Plunkett Research; 2010.

54. Wipke K, Sprik S, Kurtz J, Ramsden T. Controlled hydrogen fleet and infrastructure analysis. National Renewable Energy Laboratory. TV001; 2010.

55. Beckhaus P, Dokupil M, Heinzel A, Souzani S, Spitta C. On-board fuel cell power supply for sailing yachts. J Power Sources 2005;145:639–43.

56. Blake GD. Solid oxide fuel cell development for auxiliary power in heavy duty vehicle applications. Delphi Corporation Fuel Cells and Reformers Product Team. FC_44_Blake; 2008.

57. Brodrick CJ, Lipman TE, Farshchi M, Lutsey NP, Dwyer HA, Sperling D, et al. Evaluation of fuel cell auxiliary power units for heavy-duty diesel trucks. Transp Res Part D: Transp Environ 2002;7:303–15.

58. Agnolucci P. Prospects of fuel cell auxiliary power units in the civil markets. Int J Hydrog Energy 2007;32:4306–18.

59. Henne RH, Friedrich KA. Applications: transportation: auxiliary power units: fuel cells. In: Garche J, editor. Encyclopedia of electrochemical power sources. Amsterdam: Elsevier; 2009.

60. Lutsey N, Brodrick CJ, Sperling D, Dwyer HA. Markets for fuel cell auxiliary power units in vehicles: a preliminary assessment. Institute of Transportation Studies. UCD-ITS-RP-02-44; 2002.

61. Krumpelt M. Auxiliary power units: what, why, how, when. In: Proceedings of EU/US symposium; 2001.

62. Gaines L., Hartman C.J.B. Energy use and emissions comparison of idling reduction options for heavy-duty diesel trucks. In: Proceedings of the 88th annual meeting of the Transportation Research Board; 2009.

63. Qi Z. Application: transportation: light traction: fuel cells. In: Garche J, editor. Encyclopedia of electrochemical power sources. Amsterdam: Elsevier; 2009.

64. Dicks AL. PEM fuel cells: applications. In: Sayigh A, editor. Comprehensive renewable energy. Oxford: Elsevier; 2012.

65. Colella WG. Market prospects, design features, and performance of a fuel cell-powered scooter. J Power Sources 2000;86:255–60.

66. Lin B. Conceptual design and modeling of a fuel cell scooter for urban Asia. J Power Sources 2000;86:202–13.

67. Hwang J, Wang D, Shih N, Lai D, Chen C. Development of fuel-cell-powered electric bicycle. J Power Sources 2004;133:223–8.

68. Hwang J, Wang D, Shih N. Development of a lightweight fuel cell vehicle. J Power Sources 2005;141:108–15.

69. Wang FC, Chiang Y-S. Design and control of a PEMFC powered electric wheelchair. Int J Hydrog Energy 2012;37:11299–307.

70. Hochgraf C. Application: transportation: electric vehicles: fuel cells. In: Garche J, editor. Encyclopedia of electrochemical power sources. Amsterdam: Elsevier; 2009.

71. First Hyundai ix35 FCEV rolls off assembly line in Korea. Fuel Cells Bulletin 2013; 2013:2.

72. Zorpette G. Super charged

73. ultracapacitors.. IEEE Spectr 2005;42:32–7.

74. Pollet BG, Staffell I, Shang JL. Current status of hybrid, battery and fuel cell electric vehicles: from electrochemistry to market prospects. Electrochim Acta 2012;84:235–49.

75. Campanari S, Manzolini G, Garcia de la Iglesia F. Energy analysis of electric vehicles using batteries or fuel cells through well-to-wheel driving cycle simulations. J Power Sources 2009;186:464–77.

76. Thomas CE. Fuel cell and battery electric vehicles compared. Int J Hydrog Energy 2009;34:6005–20.

77. Nguyen T, Ward J. Well-to-wheel greenhouse gas emissions and petroleum use for mid-size light-duty vehicles. US Department of Energy. Record#1001; 2010.

78. Ahluwalia RK, Hua TQ, Peng JK. On-board and off-board performance of hydrogen storage options for light-duty vehicles. Int J Hydrog Energy 2012;37:2891–910.

79. Urban buses: alternative powertrains for Europe. Fuel Cells and Hydrogen Joint Undertaking; 2012.

80. Eudy L, Chandler K, Gikakis C. Fuel cell buses in U.S. transit fleets: current status 2012. National Renewable Energy Laboratory. NREL/TP-5600-56406; 2012.

81. Ally J, Pryor T. Life-cycle assessment of diesel, natural gas and hydrogen fuel cell bus transportation systems. J Power Sources 2007;170:401–11.

82. Saxe M, Folkesson A, Alvfors P. Energy system analysis of the fuel cell buses operated in the project: clean urban transport for Europe. Energy 2008;33: 689–711.

83. Canada gets first fuel cell bus for 2010 winter Olympics fleet. Fuel Cells Bulletin 2009;2009:2.

84. Li X, Li J, Xu L, Yang F, Hua J, Ouyang M. Performance analysis of protonexchange membrane fuel cell stacks used in Beijing urban-route buses trial project. Int J Hydrog Energy 2010;35:3841–7.

85. Neves NP, Pinto CS. Licensing a fuel cell bus and a hydrogen fueling station in Brazil. Int J Hydrog Energy 2013;38:8215–20.

86. Eudy L, Chandler K. Zero emission bay area (ZEBA) fuel cell bus demonstration: second results report. National Renewable Energy Laboratory. NREL/TP- 5600-55367; 2012.

87. Fuel Cell Application. Murdoch University. ⟨http://www.see. murdoch.edu.au/ resources/info/Applic/Fuelcells/⟩.

88. Vision receives $27m purchase order for 100 fuel cell Class 8 trucks. Fuel Cells Bulletin 2012; 2012:3.

89. Guo L, Yedavalli K, Zinger D. Design and modeling of power system for a fuel cell hybrid switcher locomotive. Energy Convers Manage 2011;52: 1406–13.

90. Miller AR, Peters J, Smith BE, Velev OA. Analysis of fuel cell hybrid locomotives. J Power Sources 2006;157:855–61.

91. Bradley TH, Moffitt BA, Mavris D, Parekh DE. Applications: transportation: aviation: fuel cells. In: Garche J, editor.

Encyclopedia of electrochemical power sources. Amsterdam: Elsevier; 2009.

92. Kim K, Kim T, Lee K, Kwon S. Fuel cell system with sodium borohydride as hydrogen source for unmanned aerial vehicles. J Power Sources 2011;196: 9069–75.

93. Boeing fuel cell plane in manned aviation first. Fuel Cells Bulletin 2008; 2008:1.

94. Fuel cells power first takeoff for DLR0 s Antares aircraft. Fuel Cells Bulletin 2009; 2009:3.

95. Uno M, Shimada T, Ariyama Y, Fukuzawa N, Noguchi D, Ogawa K, et al. Development and demonstration flight of a fuel cell system for high-altitude balloons. J Power Sources 2009;193:788–96.

96. Fuel cell project for space launch vehicles. Fuel Cells Bulletin 2002; 2002:2.

97. Sone Y, Ueno M, Kuwajima S. Fuel cell development for space applications: fuel cell system in a closed environment. J Power Sources 2004;137:269–76.

98. Sone Y, Ueno M, Naito H, Kuwajima S. One kilowatt-class fuel cell system for the aerospace applications in a micro-gravitational and closed environment. J Power Sources 2006;157:886–92.

99. First yacht with certified fuel cell propulsion. Fuel Cells Bulletin 2003; 2003:4–5.

100. First fuel cell passenger ship unveiled in Hamburg. Fuel Cells Bulletin 2008;2008:4–5.

101. Austrian collaboration unveils hydrogen fuel cell boat. Fuel Cells Bulletin 2009; 2009:5.

102. Alkaner S, Zhou P. A comparative study on life cycle analysis of molten carbon fuel cells and diesel engines for marine application. J Power Sources 2006;158:188–99.

103. Leo TJ, Durango JA, Navarro E. Exergy analysis of PEM fuel cells for marine applications. Energy 2010;35:1164–71.

104. Third fuel cell submarine handed to German navy. Fuel Cells Bulletin 2006; 2006:10.

105. Psoma A, Sattler G. Fuel cell systems for submarines: from the first idea to serial production. J Power Sources 2002;106:381–3.

106. FY 2012 progress report for the DOE hydrogen and fuel cells program. US Department of Energy Hydrogen and Fuel Cells Program. DOE/GO-102012- 3767; 2012.

107. Spendelow J, Papageorgopoulos D. Platinum group metal loading in PEMFC stacks. US Department of Energy. Record#11013; 2011.

108. Clean Energy Patent Growth Index 2012 Year in Review. Heslin Rothenberg Farley & Mesiti P.C. ⟨http://cepgi.typepad.com/heslin_rothenberg_farley_/ 2013/03/clean-energy-patent-growth-index-2011-year-in-review.html⟩.

109. Fuel cell technologies office: multi-year research, development, and demonstration plan. US Department of Energy; 2011.

110. Spendelow J, Papageorgopoulos D. Fuel cell bus targets. US Department of Energy. Record#12012; 2012.

111. Haile SM. Fuel cell materials and components. Acta Mater 2003;51: 5981–6000.

112. Farrington L. Fuel for thought on cars of the future. Sci Comput World 2003.

113. Mehta V, Cooper JS. Review and analysis of PEM fuel cell design and manufacturing. J Power Sources 2003;114:32–53.

114. Ramos-Paja CA, Romero A, Giral R, Calvente J, Martinez-Salamero L. Mathematical analysis of hybrid topologies efficiency for PEM fuel cell power systems design. Int J Electr Power Energy Syst 2010;32:1049–61.

115. Barbir F. PEM fuel cells: theory and practice. New York: Academic Press; 2005.

116. Xue XD, Cheng KWE, Sutanto D. Unified mathematical modelling of steadystate and dynamic voltage–current characteristics for PEM fuel cells. Electrochim Acta 2006;52:1135–44.

Chapter 2

The Fundamental Nature of Life as a Chemical System: the Part Played by Inorganic Elements

R.J.P Williams

Inorganic Chemistry Laboratory, University of Oxford, South Parks Road, Oxford OX1 3QR, UK

ABSTRACT

In this article we show why inorganic metal elements from the environment were an essential part of the origin of living aqueous systems of chemicals in flow. Unavoidably such systems have many closely fixed parameters, related to thermodynamic binding constants, for the interaction of the essential exchangeable inorganic metal elements with both inorganic and organic non-metal materials. The binding constants give rise to fixed free metal ion concentration profiles for different metal ions and ligands in the cytoplasm of all cells closely related to the Irving–Williams series. The amounts of bound elements depend on the organic molecules present as well as these free ion concentrations. This system must have predated coding which is probably only essential for reproductive life. Later evolution in changing chemical environments became based on the development of extra cytoplasmic compartments containing quite different energised free (and bound) element contents but in feed-back communication with the central primitive cytoplasm which changed little. Hence species multiplied late in evolution in large part due to the coupling with the altered inorganic environment.

INTRODUCTION

In this article I shall draw attention to the work of Fraústo da Silva and myself [1], [3] and [4] in an effort to demonstrate the underlying value of inorganic elements in the creation and evolution of living cells. We have to see that this value stems in part from the peculiarities of the abundances of the chemical elements in the universe, an accidental result of the Big Bang and subsequent later events in the giant stars, and then the limitations to element availability in the primitive aqueous, (sea) environment of Earth some 5×10^9 years ago. Now life started (and remains) a system of chemical activities in an enclosed volume, cells, based on large and small organic molecules and certain inorganic elements in water, Fig. 1, and one central feature is that it was in communication with a reductive initial environment. The organic molecules which initiated life were based of necessity on reduced states of C, N, O and even S due to their combination in large part with H. Since the medium for their reactions was the water of the sea, they faced fixed considerable levels of Na^+, K^+, Cl^-, Mg^{2+} and Ca^{2+} externally. The limitations of availability in the reductive, pH=8, sulphide/carbonate medium and the precipitated minerals of the primitive earth, Fig. 1, made the involvement of Na^+, K^+, Cl^-, Mg^{2+}, Ca^{2+}, HPO_4^{2-}, HCO_3^- in cells easily possible, that of Mn^{2+} and Fe^{2+} just possible, but that of other elements such as Co and Ni and especially Cu and Zn, was very restricted. It is equally clear that while the earliest reproductive life was based on a code, DNA or RNA, some system of reactions based on organic compounds and particular inorganic elements had to predate this code. A code cannot predate a system which it represents and which the system makes. It is the nature of this system which we wish to describe.

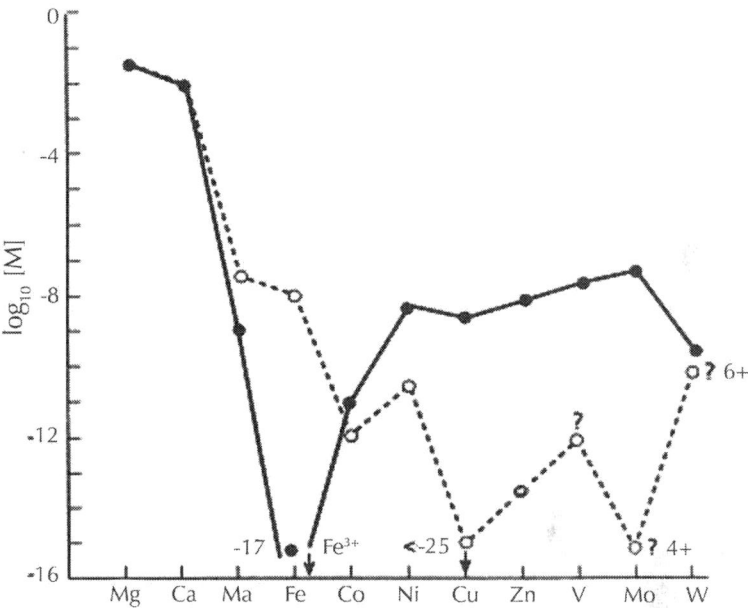

Figure 1: The free element composition of the sea: ○, in primitive times note the rough relationship to the Irving–Williams series; • today. The primitive profile is dominated by the insolubility of sulphides while the modern profile is affected by the solubility of mixed oxides and hydroxides. Note however, the special cases of the limitations on free Ca^{2+} by carbonate and the limitation of the salt content NaCl largely by the restricted abundance of chlorine.

The very nature of the system had to be based on the production from very small molecules, at the lowest level H_2O, NH_3, CO and H_2S, of polymers of the kinds $(-CH_2-)_n$, fats, $(C_6H_{12}O_6)_n$, saccharides, (amino-acids)$_n$, proteins, and (bases)$_n$, nucleotides. These polymers are unstable and require an energy input for their synthesis and we know that this input came via oxidation-reduction reactions which we shall show had to use iron catalysts in the enforced absence of any other suitable metals. In a confined space such redox reactions can lead to energy storage in gradients, for example of protons, and at some stage the proton gradient yielded ATP or simply pyrophosphate [2] in the enclosed volume or cell. Both of these compounds inevitably bound the only metal ions present in adequate concentration to bind them inside the cell, which were and are Mg^{2+} ions as we shall again show. It was pyrophosphate in conjunction with magnesium acting as a catalyst which then drove the production through condensation reactions of the polymers of all kinds. This very basic combination of reductive Fe catalysed chemistry and Mg catalyses of condensation reactions in primitive life was, as we shall show, of necessity based on these two elements alone. It remains so to this day in the cytoplasm of all cells. The difficulty in characterising life as based of necessity on such a system of chemicals is that we need to see how such a system could develop and survive in its present cytoplasmic chemical form. What we must do first therefore is to characterise the system much as we, as chemists, have characterised all other systems of matter after we know their composition. To approach the problem we shall leave to one side initially the complications of living flowing irreversible systems, in order to give a reminder of the inevitable characterisation of the nature of reversible systems on Earth and to see why we analyse them in terms of thermodynamic parameters associated with their composition. We shall then ask if we can also describe simple flow systems in terms of limitations imposed by these same thermodynamic equilibrium parameters before we return to living systems.

SINGLE MOLECULES IN SYSTEMS

Consider first a very simple molecule such as H_2O [1]. As a single molecule it is characterised by an analytical composition and a variety of properties described in all textbooks. However, as a system of molecules water has additional characteristic properties such as a bulk melting point and a boiling point. Such properties are not easily connected to the property of a single H_2O molecule.

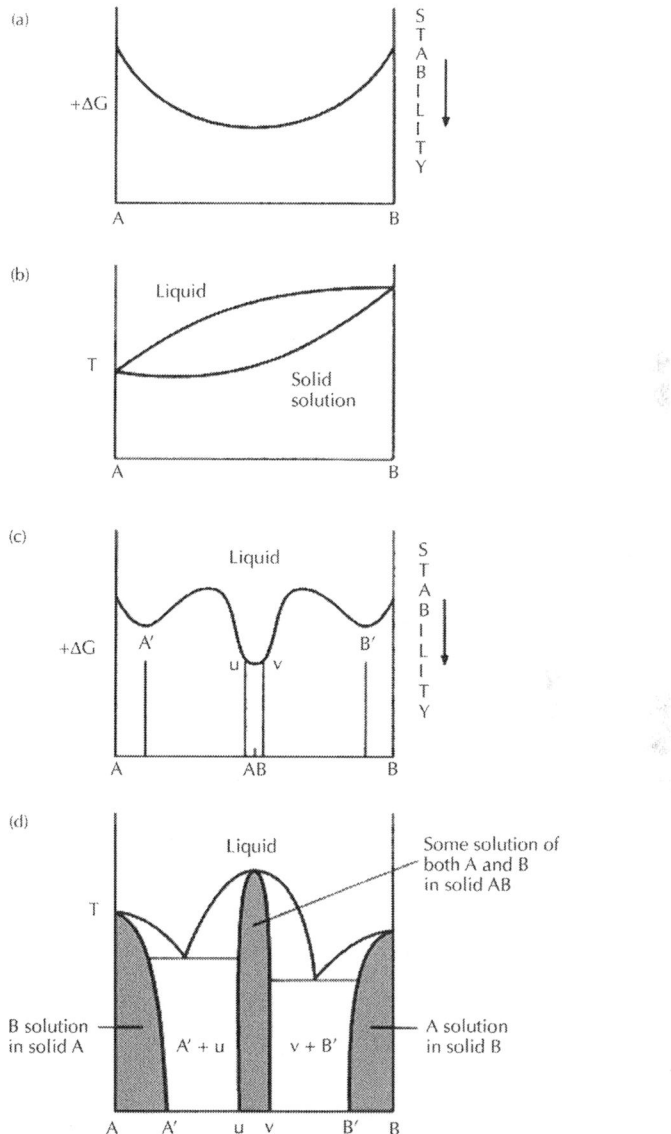

Figure 2: Free energy against composition diagrams for two substances A and B which (a) form a solution in all proportions and (c) form a com-

pound, AB, of slightly variable proportions, where u and v are the compositional limits of the compound, and two solutions from A to A' and B to B'. The corresponding phase diagrams are shown in (b) and (d). These diagrams can be compared with survival diagrams see Fig. 4.

Even more difficult is the variation with temperature of the vapour pressure over the liquid water. Instead of attempting this very difficult molecular problem we use bulk equilibrium parameters such as the dew point temperature for a given vapour pressure, to characterise the system. In other words we relate all such features of phases to thermodynamic parameters not to molecular properties. The parameters rather than atom-based descriptions then characterise the system in terms of certain free energies, heats and entropies, etc. associated with phase equilibria, Fig. 2, and phase diagrams.

The same approach can be made to larger molecules such as proteins, RNA and DNA in solution but now although the individual molecules can be described exactly by composition, even the molecules cannot be described just by simple molecular properties as was the case for single H_2O molecules. Each molecule is a small phase and has a condensation (folding) behaviour varying with temperature parallel with the thermodynamic characterisation of bulk water. Further individual polymers plus water form a two phase interactive system through surface energies and again we can parameterise such systems. In addition we can give thermodynamic equilibrium interaction constants for polymer molecules in a solution as we do for small molecules. Just as was the case in water added salts or molecules to the DNA solution alter the thermodynamic properties.

Turning now to non-thermodynamic properties associated with flow there are two obvious cases which are systems of bulk water again of the composition H_2O. They are the formation of clouds and of rivers, Fig. 3. To maintain these structures energy has to be supplied continuously and there is a vertical directional momentum imparted to the vapour from the bulk liquid while there is also an underlying structure in the cloud droplets caused by and in the surrounds of the flow. Cloud formation in particular patterns is not random but systematic. The formation of the particular structures is generated by a particular flow in a vertical temperature gradient and an air velocity in horizontal or turbulent motion. Note that vapour and droplets are in fast exchange. In the case of a river the gradients of the land together with the vertical gravitational field give rise to structure. In both cases gravity acts too on the water droplets so a set of fixed forces causes' cyclic flow within structure. These systems can form in isolated compartments so that for example we see clouds over mountains even when there are none over plains. It is important to note that they are characterised quite closely by the thermodynamic properties of water, its dew points at particular temperatures. Flow systems which are in fast exchange relative to changes of forces which constrain them are limited to a considerable degree by thermodynamic constants which help to characterise them. Let us apply the idea more generally to chemical systems.

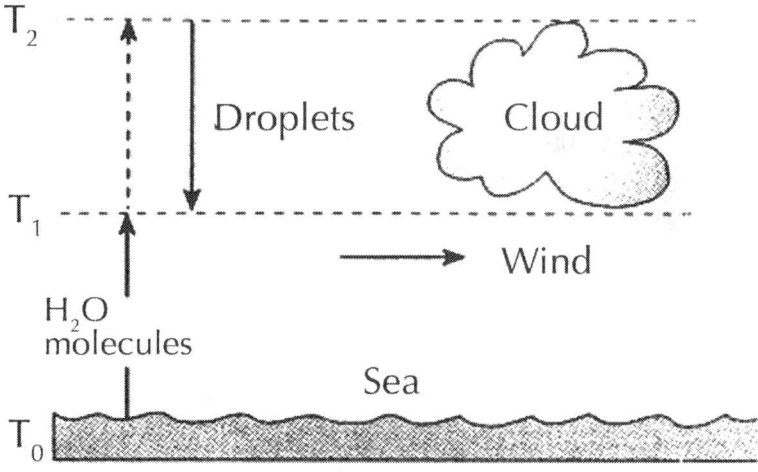

Figure 3: The creation of a flow system of given shape, a cloud, consisting of two phases, vapour and liquid. The temperature gradient, gravity and the turbulence of the wind provide the energies and force fields for the formation of the cloud. Low altitude clouds are cumulus while high altitude clouds are stratus. They are variants not species. Probably early life of prokaryotes had only variants not species.

MIXTURES OF CHEMICALS

While single substances such as water are described by their thermodynamic constants, boiling, melting and dew points, mixtures of substances are characterised additionally by composition and interactions giving combinations. Typical thermodynamic diagrams called phase diagrams illustrate regions of composition/ temperature space in which static solutions of liquids, solids or gases, and of combinations exist, maybe giving stoichiometric or non-stoichiometric compounds, Fig. 2. Simple cases are seen in the diagrams for Na (metal) $+Cl_2$ (gas) and Zn (metal) +Cu (metal) which, when mixed together at variable temperatures, give stoichiometric compounds and non-stoichiometric phases respectively. The diagrams give equilibrium zones of time-independent existence. Hence the phases in the system are characterised by equilibrium (binding) constants of the components and by melting and boiling points. We use the same logic to describe interactions between even larger numbers of components in aqueous solution. Thus we describe solutions of cupric ions plus ammonia by equilibrium constants and silver ions plus chloride ions by a solubility product. A comparative study of these binding constants while varying both metal ion and ligand was carried out in the period 1945 to 1970. This study led to classifications of series or profiles of thermodynamic constants such as the Irving–Williams series of bindings to virtually all ligands

$$(Mg^{2+}) < Mn^{2+} < Fe^{2+} < Co^{2+} < Ni^{2+} < Cu^{2+} > Zn^{2+}$$

Where the gradient in the series of constants was shown to increase in the order of ligands

O-donors (including inorganic hydroxide and oxide)

$< N$-donors $< RS^-$ donors $< S^{2-}$ (inorganic sulphide)

Moreover, these two orders applied to the variation of the solubility and the extractability into organic solvents of the ions by ligands. Other classifications of binding were described by parameters such as I/r (where I is an ionisation potential and r a radius of an ion) or by concepts of 'soft and hard' or 'b-class and a-class'. In this way a general characterisation of systems of interactions has been gained based partly on thermodynamic observations and partly on characterisation by atomic or molecular parameters. However, the more complicated the interacting units the more we use functional thermodynamic constants to characterise their behaviour. Very importantly we observe that the gradient of solubility product of sulphides along the above Irving–Williams series, (not necessarily of stoichiometric substances) is steeper than that of any known series of soluble complexes made from any ligand. Thus these solubility and stability constants have a fundamental link to the nature of the environment and any organic molecules produced in cells at all periods of the history of the Earth provided the systems approach equilibria that is they are in fast exchange. Hence we can apply them to the nature of the sea in which life originated.

In all these approaches stoichiometric bindings are considered apparently unlike the treatment of phase diagrams of mixtures of say $Cu + Zn$. In fact the two are different aspects of the same analysis – the thermodynamics of interactions. Stoichiometric combination is just the limiting result of a phase confined to a particular combination. Hence we can always plot free-energies of combination against composition to show regions of stability and instability, Fig. 4. Other dimensions such as temperature and pressure can be added to the diagrams, which illustrate multi-dimensional thermodynamic dependencies.

These thermodynamic constants of systems are used by engineers in the design of chemical plants where liquids in relative fast exchange flow even though they are only approximately valuable. Such flows generally are restricted in character by the constants of the interactions between components.

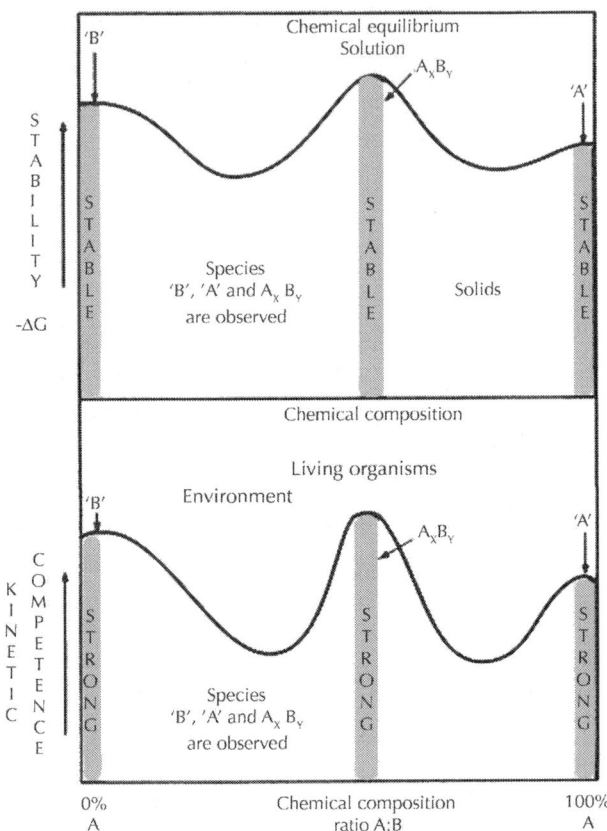

Figure 4: An extension of Fig. 2 to draw a parallel between thermodynamic stability and of kinetic stability of systems.

FLOW SYSTEMS OF MANY COMPONENTS – LIFE

When we turned from the discussion of the stationary thermodynamic properties of bulk water to the problems of its flow in rivers and clouds, we found that after noting the composition of the systems, their thermodynamic constants were still extremely useful in characterising survival (not now stability), see Fig. 4, against a background of energy input and output and within structured boundaries. Clouds and rivers are condensation phenomena and depend in fair part on dew points, even though these constants are no longer of precise descriptive value. We wish to apply this type of analysis to a many component system in flow such as life. We shall ask the question – Is a flow system such as life characterisable in fair part from knowledge of composition together with thermodynamic constants uncovered from in vitro analysis? If this is the case and we discover sets of such constants which underly the flows of chemicals in life we shall see that while a code is fundamental for the reproduction of a system it fails to reveal underlying inevitable features of 'survival' diagrams which parallel stability (phase) diagrams and which may well show the limits of compositional ranges in which species can exist, Fig. 4. (N.B. A code by itself is continuously variable within quantised limitations). These features are fundamental not just too living systems but to all chemical reactions in the universe.

ANALYSIS OF ORGANISMS

The first step to obtaining any chemical phase or survival diagram must be the analysis of its components. As inorganic chemists we shall therefore concentrate next upon the analysis of essential elements present in all forms of life. The problem is seen to be extremely complicated in that about 20 elements are essential for every known form of life which has survived or can be guessed about from fossil data. These 20 elements or a large percentage of them have to be considered as part of the underlying character of life no matter how the system is structured, no matter what energy is put into it, nor indeed how it is reproduced. There is only one general form of life known to us. Let us put to one side for the moment the elements required to build large molecules such as proteins and DNA, Fig. 5, that is all the non-metals H, C, N, O, P, S and Se, and turn our attention to the metals present in cells, for example Na, K, Mg, Ca, Mn, Fe, Co, Ni, Cu, and Zn asking if all are essential for all forms of life and then going on to the question as to why they are present in the concentrations observed.

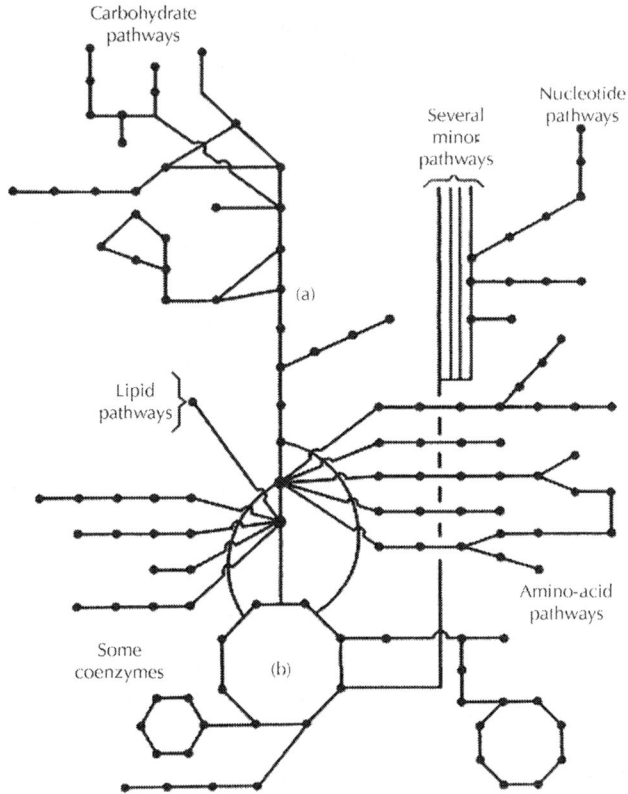

Figure 5: A schematic integrated circuit of substrate, C, H, N, O, distribution which must be in flow balance so that each set of products receives the correct amount of the elements and the correct amount of energy by feedback regulation. Each point represents an enzyme and the track (a) could be the glycolytic pathway while track (b) is the citric acid cycle. From any one pathway there has to be feed-back to others using common coenzymes, e.g. NADH and ATP, so that the whole maintains homeostasis or growth in a pattern of chemicals (after Kauffman).

Now we have already stated that life is not an isolated cellular system but is and was always in communication with its environment from which it takes and to which it gives both chemicals and energy. We have characterised the major part of the primitive environments, the early sea, as reductive and we have noted the major thermodynamic solubility products which limited it, e.g. for hydroxides, carbonates, and sulphides, Fig. 1. Hence one part of the system (the external part) is well recognised quantitatively, that is in concentrations. Although it is a system in flow we can characterise it quite closely by thermodynamic constants in various gradients. Its elements are in fast exchange compared with the time scale of evolution of Earth or life. We must now turn to the value of these chemicals to the survival of the internal part (the enclosed cellular space) of the system. We are concerned with the characterisation of this two part system since we wish to discover why survival of organisms is confined to certain compositions to this day much as we see for stability in phase diagrams, Fig. 4.

FUNCTIONAL VALUE

We see that a living system of molecules can only be maintained if certain chemical elements are present in solution and therefore they must be needed in certain functional ways related to their concentrations. The control over the concentration requires selected use of energy to different element gradients. For example for the ability to form the required fatty substances to give a membrane, sources of C and H but also of C/H chemistry are needed, and in order to form polymers in aqueous solution, sources of C/H/N/O/S/P availability are required and there is a need to devise their syntheses. Energy was always also required to incorporate these elements from simple molecules into larger ones. Again these combinations can be maintainable in water as separate membranes and soluble molecules only if they are for the most part negatively charged. Three problems therefore arise: (1) the system of the sea (the original environment of life) and the enclosed volume is not stable unless NaCl, the dominant osmotic salt in the sea, is removed from the enclosed volume since it also includes a considerable concentration of organic molecules. (2) The necessary rejection of NaCl still leaves the problem of the unstable negative charge balance in the enclosed space. It is observed that to achieve this balance potassium ions are taken up into all cells. (3) The synthesis of large molecules demands energy and catalysts. Thus cellular life is characterised not just by organic chemicals but firstly by Na^+, K^+ and Cl^- concentrations used to control ionic strength, osmotic pressure and rough charge neutrality in all cells. They and only they were and are so used since they unlike other ions were and are all available and soluble in the required concentration range 10^{-1} to 10^{-3} M in the early sea and they do not interact with organic matter. In cells they are in somewhat energised gradients relative to the sea but otherwise the cytoplasm reflects limitations of the properties of simple salts as well as their abundances. Except for a few situations we do not need to characterise their combinations except by availability and the small required energy inputs. These ions are always in fast exchange in any compartment. Strange as it may seem whatever characterised the primitive sea went a long way to characterise inevitable features of living systems, cells: Na^+, K^+ and Cl^- perform essential functions for life in quite fast exchange. This means that their pumps at least are characterised by thermodynamic binding constants (inside and outside) which reflect the two compartments, sea and cytoplasm.

Let us consider next the elements Mg^{2+} and Ca^{2+}. Again the sea is rich in these elements ($>10^{-3}$ M) while all cellular living systems hold the free ion levels of Mg^{2+} at 10^{-3} M and Ca^{2+} at 10^{-6} M or lower in the cytoplasm. Note again that these two ions are always in fast exchange so that binding constants to pumps reflect their concentration gradients of necessity. Why are these elements at these concentrations essential features of life? The rejection of calcium from cells was necessary since many inorganic and organic anions form insoluble salts with this cation. Mg^{2+} on the other hand forms soluble complexes with many of the anions but not precipitates. Such complexes are necessary in fact to organise and activate the organic molecules for hydrolysis and condensation and Mg^{2+} is used as life's very dominant weak Lewis acid. Why? Now we have already noted that the very nature of life the synthesis, condensation of small molecules, and degradation of polymers requires energy and catalysis. We have also seen from model studies that the binding of metal ions to oxygen donors, which are the basic centres of many small substrates and to ATP, the source of energy, is weak, generally less than $10^6 M^{-1}$ for oxalate and pyrophosphate for example. In the inevitable presence of sulphide only Mg^{2+} could form complexes with such oxygen-donors. It was an inevitable consequence of the nature of the geochemistry of the primitive sea that the major Lewis acid catalyst for the original cellular system was magnesium. This is selection by nothing other than thermodynamic necessity not 'biological fitness'. We see too that once the observed organic chemical system was established it was not changeable and had to be protected. Thus just as the sea is characterised by the limitations imposed by binding constants of Mg^{2+} and Ca^{2+} to carbonate so is life characterised by their binding constants to internal organic chemicals. For the system to operate it is essential that the binding constants for the inside face of the pumps, of any carriers (chaperones), of transcription factors controlling production of binding agents, and of active sites for a given fast exchanging metal ion must be closely equal otherwise they all cannot be occupied usefully. The constants applicable to the metal complexes and solubility products of the sulphide-containing primitive sea were also a necessary feature of early life, removing many metals, since the living system is one, sea environment plus enclosed cellular volume, with energy supplied to assist survival of the enclosed volume.

REDUCTION

As stated all cells need an energised reducing internal cytoplasm, which is necessary to generate the monomers such as amino-acids, fats, sugars and nucleotides for the construction of polymers. As we have seen also the primitive environment, the sea, it was in fact reducing and there were available CO (CH_4), NH_3, H_2, H_2O and H_2S. Now the redox reactions of these molecules to make larger molecules, e.g. $CO+H_2 \rightarrow HCHO$, require a reducing transition metal catalyst. As stated the transition element availability in this atmosphere would have approximately followed the solubility products of sulphides, i.e. $Mn^{2+} > Fe^{2+} > Co^{2+} > Ni^{2+} > Cu^{2+}$ (Cu^+) $< Zn^{2+}$, i.e. in the Irving–Williams series and where $Mn^{2+} > 10^{-6}$ M and

Cu^{2+} (Cu^+) $<10^{-20}$ M. Given these numbers it is clear that while a cell needed at least one redox active metal ion in its cytoplasm as a catalyst it had little opportunity to obtain any but Mn^{2+} or Fe^{2+} since the gradients of binding constants of internal organic ligands are smaller than that of external inorganic sulphide in the Irving–Williams series. Since manganese does not have suitable redox potentials below 0.0 V in a sulphide medium, iron alone was of value with a limited concentration of about 10^{-7} M, Fig. 1. The cell is seen to be a part of a compartmentalised system which has an essentialinside/outside disposition of elements dependent on binding constants whence only Fe^{2+} could be used. Again all its binding constants to organic binding sites in the cytoplasm must be closely equal at 10^7 M^{-1}. In fact the differences between the primitive sea and the primitive enclosed cellular contents of free inorganic ions was probably quite small since sulphide would diffuse freely, like CO_2, into the cell. As far as inorganic chemistry is concerned we can see again that universal necessity was the mother of the invention of life. Thus life is characterised by the thermodynamic constants of the whole system of the primitive sea plus the internal synthetic organic chemistry which was made possible by trapping energy and anions inside an enclosed space. Very few elements, but note especially calcium, are far removed in free ion concentration in the cytoplasm from the values controlled by inorganic thermodynamics of the primitive sea. Here again we note the very low levels of free Co, Ni, Cu, Zn, and Mo (W) in cells. This sea therefore characterised life together with the thermodynamics of binding (stability constants) to organic ligands in the enclosed space which became the cytoplasm. An apparent triumph of this system in the cytoplasm is that it has remained roughly constant throughout evolution even though the environment has changed grossly. This may be a necessity however since the basic organic chemistry in the cytoplasm could not be changed, Fig. 5. It is based on energised reductive metabolism and mild acid catalysis to give lipids, proteins, DNA (RNA) and polysaccharides which requires approximately the same inorganic element free ion concentrations in the cytoplasm as a chemical necessity no matter that it is coded today. We have already seen that energisation, like catalysis of metabolism, is based largely upon Fe^{2+} and Mg^{2+} chemistry.

Without elaborating these principles further we can state that quite independent of any code for the management of life's chemical systems a certain profile of free element concentration, Fig. 6, was and is demanded in the cytoplasm for the survival of all cells. Slight variations on the theme are allowable only. Thus life's cytoplasmic chemistry exists within a narrow range of thermodynamic binding parameters forced in large part opposite the primitive environmentally available elements. (The parallel is with clouds of similar composition (droplets) but of different shapes at different altitudes but all related to the dew-point thermodynamic properties of water). We shall have to consider how rather different species could arise later.

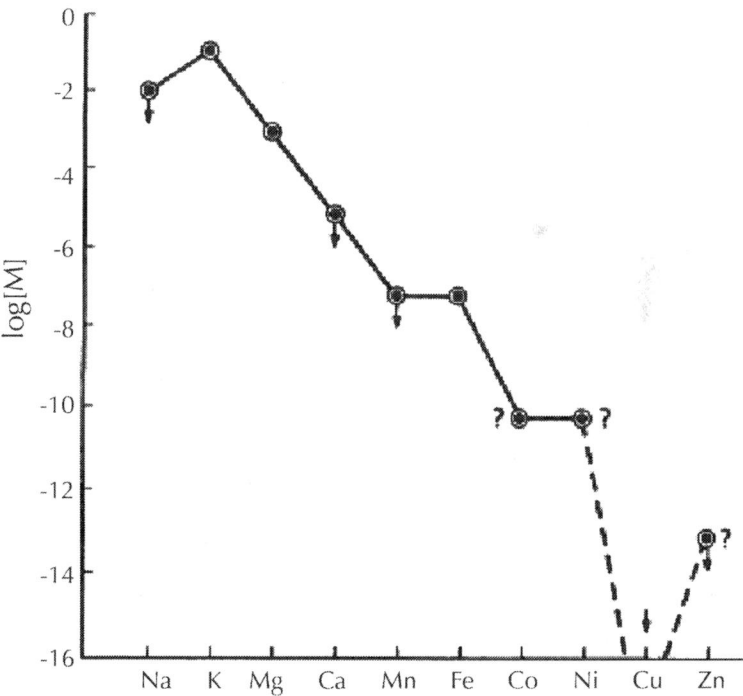

Figure 6: The profile of the concentrations of the free metal ions in the cytoplasm of the most primitive cell. Note the parallel with the profile

of the primitive sea, Fig. 1, and with the inverse of the Irving–Williams stability series. The ions of the metals are all M^{2+} except for Cu^+, K^+ and Na^+. A parallel profile exists for some anions, e.g. Cl^-, HPO_4^{2-} and HCO_3^-. The figure is based on a pH=7 and an H_2S concentration of 10^{-3} M.

Extending the picture to other elements we can give therefore a generalised quantitative picture of the free metal ion content of the cytoplasm of all cells. There is a pattern of free ions therefore in the cytoplasm which is in large part parallel with that of the initial sulphide environment but with a lower gradient and follows closely the Irving–Williams series. We shall call the profile of free element concentrations in Fig. 6 the free metallome. Of course a full understanding of the way in which a primitive or a modern cell incorporates these principles can only be appreciated from a parallel study of the use of non-metals, of the way the system avoids poisons, and of the employment of energy to maintain the requisite concentrations and structures. The coming study of the proteome must reveal some of these considerations.

THE SECOND STEP OF INORGANIC CHEMICAL EVOLUTION IN LIFE

In the next step by which primitive systems could evolve, the effective number of useful elements was increased by trapping some of them in irreversible sites relative to the free ion. This liberates them from the constraints of the free metallome but new demands on binding occurred as well as new syntheses. A code was probably essential at this stage. We have to examine the incorporation of iron in heme, cobalt in vitamin B_{12}, nickel in F-430 and magnesium in chlorophyll in prokaryotes. Effectively these four complexes and molybdenum (or tungsten) in their cofactors are new 'elements' in biological systems in their own kinetic traps. Each of them must have transport and transcription-limited properties but they must be handled other than by known empirical roles of equilibrium binding strengths for free ions. Once an element is incorporated into an irreversible non-exchanging unit, the unit's binding constant represents a new fixed constraint and its introduction immediately allowed diversity of speciation to the degree that eventually plants (chlorophyll) and animals (no chlorophyll) became separate systems. The introduction of irreversible sites in any co-enzyme like these metal chelates means that the total concentration of a given metal in the cytoplasm or in cells generally is not related to the observed free ion concentration which is a property of the primitive cytoplasmic cell system. Here we need to consider the controls over synthesis of organic material limited not only by the transportable free metal in the cell but by limitations on synthesis, see below.

THE COMING OF DIOXYGEN

Dioxygen, or its related H_2O_2 and O^{-}_{2}, was generated by simple prokaryote cells from water due to their demand for hydrogen. The by product, the production of cholesterol by dioxygen reactions, brought about the second big change in evolution – the appearance of multi-compartmental cells due to spatial separation – the eukaryotes. We cannot go into detail here but effectively both a larger cell and a diversified cell came about through changes in membrane composition. With compartmentalisation came the need for new communication between the cytoplasm and the new compartments to increase survival value. It is here that the calcium gradients already present in primitive prokaryote systems were increased and pressed into service. To signal external events and generate an internal response calcium was allowed to enter the cytoplasm temporally. A new set of ligands all of the same binding constant of 10^7 M^{-1} was synthesised including pumps, transport proteins (chaperones), triggering and transcription activities. This equilibrium constant then characterises the eukaryote cytoplasm in a somewhat novel way and it is a value that the liganding entities had to find no matter how the ligands are coded. It is a system requirement for the use of calcium as a sensitive trigger, Fig. 7. Note that the new compartments, vesicles with inside/out membranes relative to the outer membranes, contain 10^{-3} M Ca^{2+} and as they are like the sea, they as well as the external face of the cell have proteins all of binding constant around 10^3 M^{-1}.

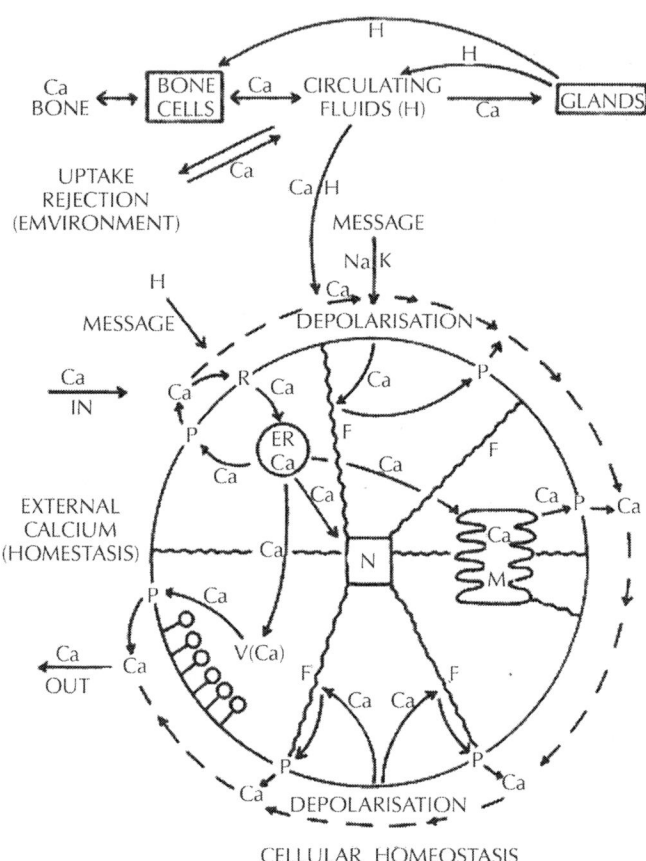

Figure 7: The network of Ca^{2+} flows in a modern organism linking the external, the vesicular and the internal cytoplasmic solutions, where H is a hormone; P is a pump; F, a filament; N, the nucleus; M, a mitochondrion; V, a vacuole; ER the endoplasmic reticulum; and R is a receptor. Binding constants in the cytoplasm are close to 10^7 M^{-1} while externally and in vesicles they are 10^3 M^{-1}.

Now dioxygen changed the possible environmental components of the living system, see plus cells, once sulphide was removed, Fig. 8. Thus iron became difficult to obtain and new modes of uptake of the element had to be found. Unlike the novelty of the calcium binding, which is a short-lived invasion and did not affect the cytoplasmic system, the new iron binding had to be such that it could provide the cytoplasm with the required $10^{-7}M$ Fe^{2+}. The ways in which this was achieved are well documented but the significance of the requirement throughout evolution of 10^{-7} M Fe^{2+} must not be missed. It, like 10^{-3} M Mg^{2+}, 10^{-1} M K^+ and so on, is a fundamental feature of life related to the nature of the primitive sea. Of course it is related to to the need to maintain a fixed pH, a fixed redox range from about -0.3 V (NADH/NAD) to -0.0 V (RSH/(RS)$_2$) in the cytoplasm, and a fixed organic chemical metabolic pattern, Fig. 4.

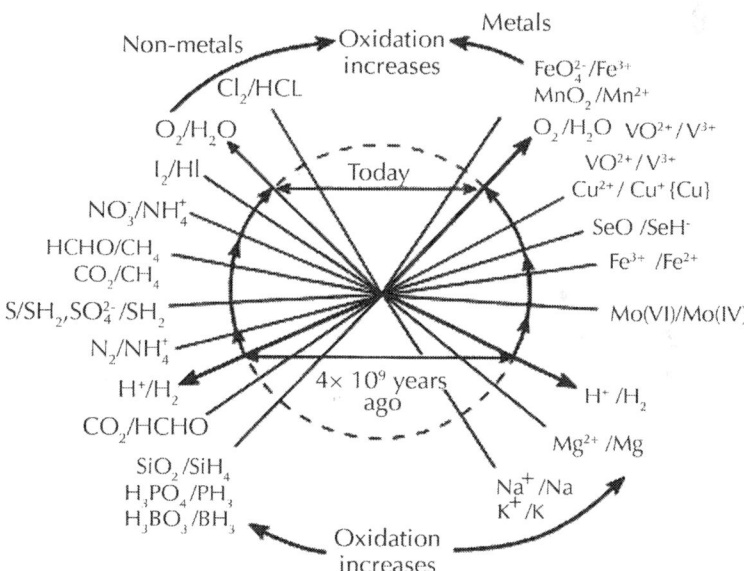

Figure 8: The changes in oxidation conditions with time introduces new species to the environment. 4×10^9 years ago the redox potential of the sea was close to -0.2 V vs. H_2/H^+ electrode at pH=7. Today the redox potential of the aerated sea is nearer to $+0.8$ V on the same scale.

The environment (the sea) had changed dramatically in the opposite sense from Fe^{2+} for the elements Cu^{2+} and Zn^{2+} which were released from sulphides and now had considerable concentrations

~10^{-6} M, Fig. 1. These elements were dangerous poisons for the cytoplasmic system and had to be controlled in the enclosed volume at concentrations below 10^{-10} M (zinc) and 10^{-15} M (copper). In this article it is not important to look at the way this was managed by the production of new ligands with feed-back controls over the concentrations of these two elements as free ions. The important observation is that all forms of life had to maintain these very low levels in the cytoplasm. They do so by generating series of ligands, pumps, carriers, buffers, transcription factors all having very closely similar binding constants for a given element of necessity, and all of which can be related to the free ion levels which must not be exceeded, Fig. 9. The necessity is the maintenance of the cytoplasmic system and it is again not important here that this became coded. Just as in the case of calcium so in the cases of copper and zinc so long as the concentrations of free zinc and copper were maintained at levels close to those of Fig. 9, to protect the cytoplasmic Mg^{2+}/Fe^{2+} primitive system and thence all of metabolism it was now possible to use the elements such as Cu and Zn in irreversible combined forms even in the cytoplasm, e.g. in superoxide dismutase but especially in new compartments. Zinc could even be allowed in cells to 10^{-11} M in reversible exchange so that it could participate in signalling to DNA in Zn fingers. (Reversible exchange does not have to be mediated by free ions but pump faces, carriers and transcription factors must have closely equal binding constants). Moreover in compartments such as vesicles or on the outside of membranes much greater use could be made of the elements inoxidising conditions and not so restricted in concentration or metabolism. Thus copper and zinc are generally deployed in many enzymes outside the cytoplasm. It was the development of the differential use of elements in different compartments in single eukaryote cells which allowed the expansion of species with special chemistries confined in special spaces, Fig. 10, and then to the evolution of multi-cellular organisms.

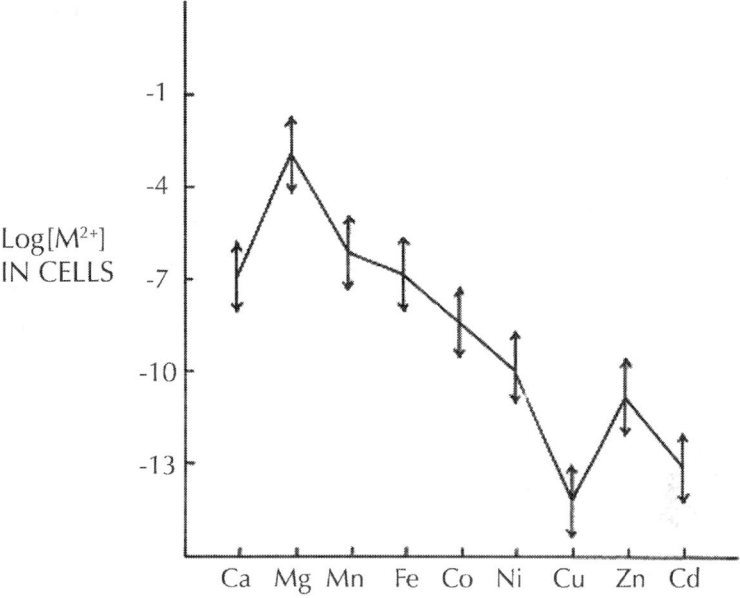

Figure 9: The suggested profile of the free element content of the eukaryote organisms, its free metallome, after the coming of dioxygen. Downward arrows show outward pumps.

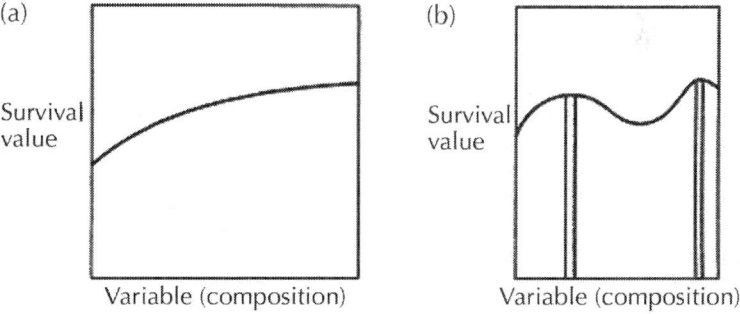

Figure 10: (a) The continuous variation of primitive living systems, no species, compared with the formation of more recent species (b) where species are identifiably separated by differences in composition. Note that composition varies along some 20 axes.

MULTI-CELLULAR ORGANISMS AND INORGANIC CHEMISTRY

The requirement for multi-cellular organisms is to use external space in a controlled way. This has been achieved by generating relatively fixed disposition of cells using connective tissue and new modes of communication between cells. It is not enough for cells to stick together in colonies. We are not concerned here with the changes of code which were brought about to secure this. Our attention is upon the fact that oxidising conditions were required for the chemistry of synthesis of cross-linked connective tissue and also for much of the production of new organic chemicals for signalling. Here we find the value of copper and zinc to dominate that of iron and magnesium. Copper is a very fine element in oxidases external to the cytoplasm where heme iron, but not iron itself, can also be used. Zinc is the major hydrolytic element outside the cytoplasm not magnesium. Both zinc and copper enzymes are used to produce or hydrolyse connective tissue and to make a great variety of inter-cell messengers. None of this chemistry disturbs the cytoplasmic activities of Fe^{2+} and Mg^{2+} inherited inevitably from the primitive system. (The new zinc finger etc. exchanges in the cytoplasm are a screened system of Zn^{2+} effectively $<10^{-11}$ M when it does not damage Fe^{2+} or Mg^{2+} bindings). Evolution is then confined by the necessities of systems involving a variety of binding constants now in both reducing and oxidising conditions constants in different compartments. In no way does this limit the number of species since the combinations of different kinds of compartments all obeying the same rules but cleverly different in organic chemistry is unlimited. The composition of organisms of a given species will be restricted to certain zones with the cytoplasm of all cells of all species having very similar free element concentrations but very variable vesicle and membrane contents. Our books [1],[3] and [4] elaborate these principles which chemistry (mainly inorganic) impose inevitably on exchanging flow systems which are seen in living organisms in given environments.

ORGANIC MOLECULES AND LIGANDS IN CELLS

In the above, no description has been given of the limited production of ligands, which includes even proteins and DNA in cells, although any description of ligand binding must include the ligand concentration. The problem here is not so much related to the uptake of elements or to the energy employed in incorporating them but in achieving a balanced integrated system of synthetic activities. In the initial uncoded system this could not have operated strongly, though clearly in any system which only exists if many components are present the system, fails if any components are deficient and maybe if any are in excess. To secure correct balanced expression needs feed-back to increase the presence of those components in low concentration, while stopping production of those in danger of being produced excessively. To this end the production of a component is managed by stimulation when there is deficiency, while reducing production of other components which would otherwise appear in excess. This mutually interactive network of synthetic activity regulated balanced production of amino-acids, bases, lipids and saccharides, all of which rely on the same source of energy and of the elements C, H, N and O, Fig. 5. Thus we observe a consented switch on or off of all activity when supplies are low. In the case of components which have a metal ion incorporated, reversibly or irreversibly in an organic chelate, then these elements must be part of the feed-back, feed-forward network of the synthesis of this chelating agent. Thus we observe that at certain low levels of say $[Mg^{2+}]$ or $[Fe^{2+}]$ in all cells all organic synthesis (and energy capture) ceases. Hence there is a connection, which is known between free levels of several inorganic elements and the cellular synthesis networks via binding proteins. This applies to ring chelate production so that there is a link between for example heme and free iron, which may be different from that which controls free iron itself in concentration dependence. This area of investigation is closely related to the study of the proteome.

SUMMARY

There is an over-riding need to view organisms in a systems (thermodynamic) style, Fig. 11, not in an isolated molecule style. This draws attention away from the qualitative coded nature of life to the underlying problems of quantitatively managed activity. The activity is connected to the environment surrounding the enclosed cellular spaces both in the need for sources of material and energy. It is easier to see the essence of the system through the study of the inorganic content, free and bound elements, which we call the metallome, than through the organic content, the proteome, since these elements are often in treatable exchange limited by thermodynamic factors. It is then not possible to appreciate the nature of systems through the study of DNA sequences, the genome, much though the environment, the metallome, the proteome and the genome are connected by feedback circuitry. The underlying importance of the inorganic chemistry of living systems must be recognised in all biological studies.

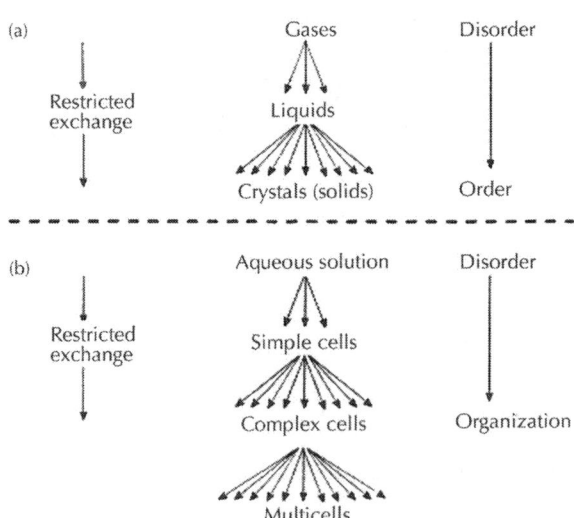

Figure 11: An illustrative diagram of the evolution of order (structure) and organisation (flow systems) in the universe.

NOTE ADDED IN PROOF

After this paper was submitted we became aware of recent papers by C.E. Outten, T.V. O'Halloran, Science 292 (2001) 2488 and by S.C. Burdatte, S.J. Lippard, Coordination Chemistry Reviews 216–217 (2001) 333 and see R.J.P. Williams, Coordination Chemistry Reviews 216–217 (2001) 537 giving further details of free ion, especially Zn^{2+}, concentrations in cells. There is general agreement that free Zn^{2+} concentration is well below 10^{-10} M. However the significance of such low concentrations, possibly less than one ion in an enclosed biological microvolume, is confused in the writings of these authors. The problem is not new and has been thoroughly analysed over many years, see H. Bauer, J. Scientific Exploration 4 (1990) 49. It is entirely respectable and realistic to refer to the free ion concentration even when it is almost vanishingly small. Inorganic chemists are used to solubility products of $<10^{-30}$ and stability constants $>10^{30}$.

ACKNOWLEDGMENTS

Although I am sole author of this paper by invitation it is important to recognise that the ideas in it have been developed jointly with Professor J.J.R. Fraústo da Silva, see the references below.

REFERENCES

1.   J.J.R. Frausto da Silva, R.J.P. Williams, The Biological Chemistry of ´ the Elements, 2nd Edition, Oxford University Press, Oxford, 2001.

2.   R.J.P. Williams, J. Theoret. Biol. 1 (1961) 1–13.

3.   R.J.P. Williams, J.J.R. Frausto da Silva, The Natural Selection of ´The Chemical Elements, Oxford University Press, Oxford, 1996.

4.   R.J.P. Williams, J.J.R. Frausto da Silva, Bringing Chemistry to Life, ´ Fig. 11. An illustrative diagram of the evolution of order (structure) and Oxford University Press, Oxford, 1999

Implementation and Analysis of a Chemical Engineering Fundamentals Concept Inventory (CEFCI)

Y. Ngothai and M.C. Davis

School of Chemical Engineering, The University of Adelaide, Adelaide, SA 5005, Australia

ABSTRACT

An effective understanding of fundamental concepts in Chemical Engineering can have an enduring affect on the ability of students to achieve success in their degree. Concept inventories are tools implemented to analyse students understanding of the fundamental concepts in their learning programs. A study at a large University in Australia has facilitated the development and implementation of a Chemical Engineering Fundamentals Concept Inventory (CEFCI).

This concept inventory provides a quantitative means to predict areas in which course development can be focused. The purpose of this paper is to illustrate the results from the CEFCI implemented at our institution, which follows similar research at the University of Melbourne (Shallcross, 2010). An outline of the development of the CEFCI questions is provided, showing the thorough methodology implemented to ensure a strong foundation for the CEFCI. Results from implementation of the CEFCI, along with the implications and limitations of these results are provided. Unlike previous research, rigorous analysis of the results through implementation of statistical methods has been completed. This provided a novel approach through which to analyse the effectiveness of both the inventory and the teaching of foundational concepts in the School of Chemical Engineering. The results of the implementation and analysis of the inventory displayed areas for constructive development in areas of synthesis and instruction of key concepts. Furthermore we believe a longitudinal study will facilitate improved understanding and implementation of the CEFCI. A similar tool could be utilised for other engineering disciplines, providing broad appeal for this current research.

INTRODUCTION

Concept inventories can be powerful tools for analysing an individual's conceptual understanding of fundamental concepts that underpin their core knowledge. The School of Chemical Engineering has developed a CEFCI that is designed to establish the misconceptions that students posses regarding the foundational concepts appropriate to their Chemical Engineering education.

A sound understanding of foundational principles can form a strong basis on which further knowledge is built. The importance of concept inventories is demonstrated through similar tools developed for Thermodynamics, Fluid Mechanics and Heat Transfer (Jacobi et al., 2003, Martin et al., 2004, Krause et al., 2004, Evans et al., 2003 and Midkiff et al., 2001). However the students understanding of the fundamental concepts that form the foundation of Chemical

Engineering have had some consideration, however statistical analysis and in depth development has, to date, not been completed (Shallcross, 2010). The CEFCI implemented and developed by coupling the results with rigorous analysis techniques facilitates a novel new tool to allow instructors within Chemical Engineering to assess and implement subsequent teaching changes to assist improved understanding.

The CEFCI was initially administered to students sitting Chemical Process Principles II, their second fundamental course in Chemical Engineering. The results analysed are from the 2007 and 2008 classes that undertook the course. Underlying understanding of fundamental concepts was analysed in the CEFCIresults, with consideration of knowledge retained from the students' first fundamental Chemical Engineering course. Improved understanding of the commonly held misconceptions that students possess in the Chemical Engineering discipline was facilitated through implementation of the CEFCI (Streveler et al., 2006 and Olds et al., 2004). Statistical analysis techniques were implemented to provide greater understanding of the effectiveness of the instrument and facilitate inventory development based on a quantitative understanding of the inventory. The structure of the inventory is subsequently considered with a discussion of components of the inventory and how these provide improved benefits in the implementation and analysis of the inventory. Finally future work that can be undertaken to enhance the effectiveness of the current concept inventory is outlined.

An illustration of the application of the CEFCI at our engineering institution is provided for the first time. In depth analysis demonstrates the validity of the instrument whilst outlining areas for continual improvement opportunities.

METHODOLOGY

The array of questions for the CEFCI were developed and sourced so that they correlated with the concepts taught in Process Systems and Process Engineering I. Elementary Principles of Chemical Engineering (Felder & Rousseau, 2005) formed the foundation on which many questions were developed due to its use as the primary course text book. An understanding of students misconceptions through critical discussion and focus sessions, facilitated scaling of the final concept inventory to the current 20 question format.

Analysis of the course coupled with a concurrent development of questions facilitated the development of problems that were designed to reflect the fundamental concepts taught in the Process Systems and Process Engineering I course.

Experience of previous misunderstandings held by students along with focus sessions revealed primary areas of student misconceptions. These procedures aided in the development of questions to address these areas of misconception. 'Distracters', which were multiple choice answers that reflect common misconceptions associated with these questions were generally incorporated into these questions. Incorporation of these questions and distracters were deemed crucial to gain improved understanding of the extent of misunderstandings in these crucial concept areas.

Final analysis and focus discussions removed questions deemed to be repetitive and ambiguity in questions was assessed by a panel of students reducing the inventory to the final 25 questions reflecting the current inventory structure.

The concept inventory was administered in two designated lectures, in the first and again in the final week of the semester. The students were provided with the questions and given an hour to complete them. On completion the students' answers were returned to the lecturer, along with the question paper. This methodology ensures that results bias is minimised in subsequent years from students potentially distributing the questions to students in future classes.

Students are not obligated to sit the concept inventory, however participation is encouraged. As an incentive for undertaking the concept inventory, students are provided with full marks in one assignment. This upgraded mark constitutes a percentage of their overall grade. Allocating assessment marks has seen greater than 95% of students, of all abilities participate in the CEFCI. This has provided unbiased data that is representative of the complete range of student abilities in the class.

To date, the results provided represent testing of 140 students with 280 individual papers having been completed. To ensure consistency only the results for those students that sat both the pre-course and post course concept inventory are presented.

INVENTORY STRUCTURE

The questions used in the CEFCI can be readily broken down into core concept areas. The relevant questions and key concept areas are as follows. The full inventory is outlined in Appendix A.

- Questions 8 analysed students comprehension of the conversion of measurement units (CGS and wt.% vs. mol%). Disastrous events with enduring consequences, as a result of unit conversion errors, are well documented (Nelson, 1997).

- Questions 11, 12, 13, 17, and 18 considered the conservation of mass principle. These problem statements reflect the conservation of mass principle in its most general form, along with adaptations which apply the principle to open and closed systems, systems undergoing chemical reactions and reacting systems containing limiting species. As a foundational physical and chemical principle understanding was deemed paramount.

- Questions 6, 7, 10, 16 and 20 require the students to elaborate on the outcome of a change in energy of a simple unit process with or without a change in the state of matter of the system. Appropriate knowledge of energy concepts is crucial for design, maintenance and operation of many industrial processes.

- Questions 1, 3, 5 and 14 focussed on the nature and influence of with-in system parameters (pressure and temperature) on the gaseous state unit operations. Gaseous state unit operations are common in the process industries, requiring an appropriate level of understanding of the influence of system parameters on their behaviour.

- Question 2 required consideration of the influence of mixing on the molar composition of a solution. Complex mixing processes are a foundation of Chemical Engineering and are therefore of importance.

- Question 4, 9, 15 and 19 questioned fundamental energy concepts. Practical consideration was given to the underlying kinetic, potential, latent and sensible energy change processes for closed system processes.

RESULTS

The results presented are from the 2007 and the 2008 Chemical Process Principles II classes. Fig. 1 and Fig. 2 provide a convenient comparison between the pre-course and post course results on theCEFCI.

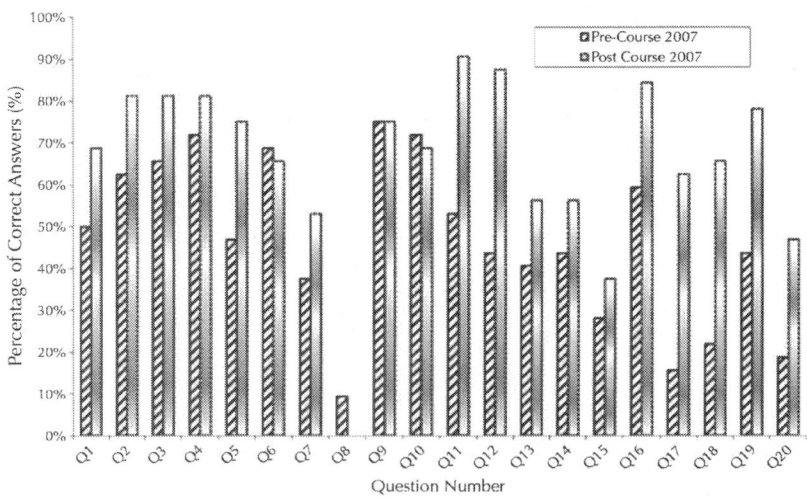

Figure 1: Concept inventory pre-course and post course results for the Chemical Process Principles II class of 2007.

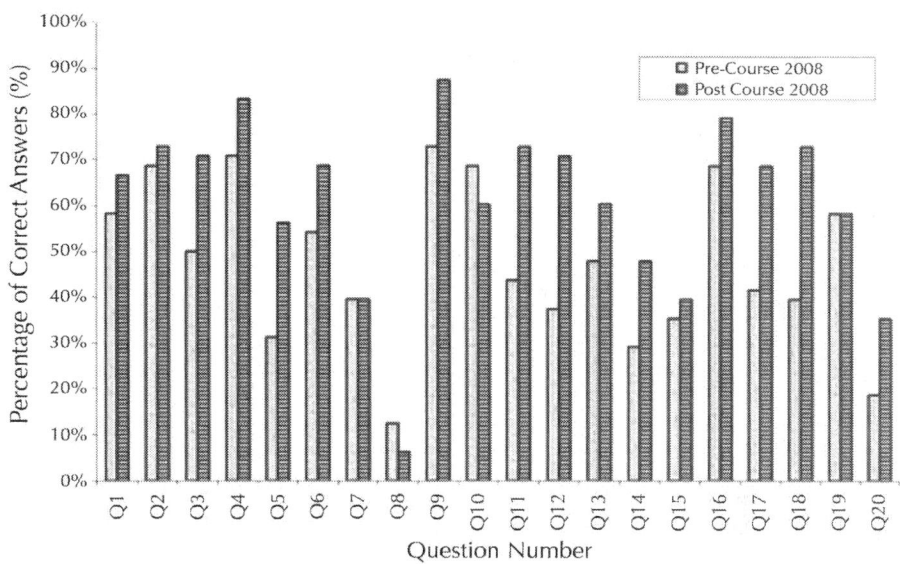

Figure 2: Concept inventory pre-course and post course results for the Chemical Process Principles II class of 2008.

Fig. 1 shows a number of interesting results. It can be seen that there is a decrease of 10%, in the number of correct responses to question 8, indicating a lack of understanding of this question, and suggesting guessing in the pre course inventory. In contrast there is a considerable increase in the number of correct responses (>20%) for questions 1, 2, 7, 11, 12, 16, 17, 18 and 19 suggesting these concepts were well understood.

Similar to Fig. 1, Fig. 2 shows that the 2008 cohort had a decrease in the number of correct responses for question 8. However, compared to the 2007 cohort, there were a lower number of questions with a greater than 20% improvement, questions 5, 11, 12, 17, 18, 20.

Cross yearly comparison highlights the difference in student performances from the two different cohorts tested. Fig. 3 and Fig. 4 provide a comparison respectively, of pre-course and post-course results from 2007 to 2008.

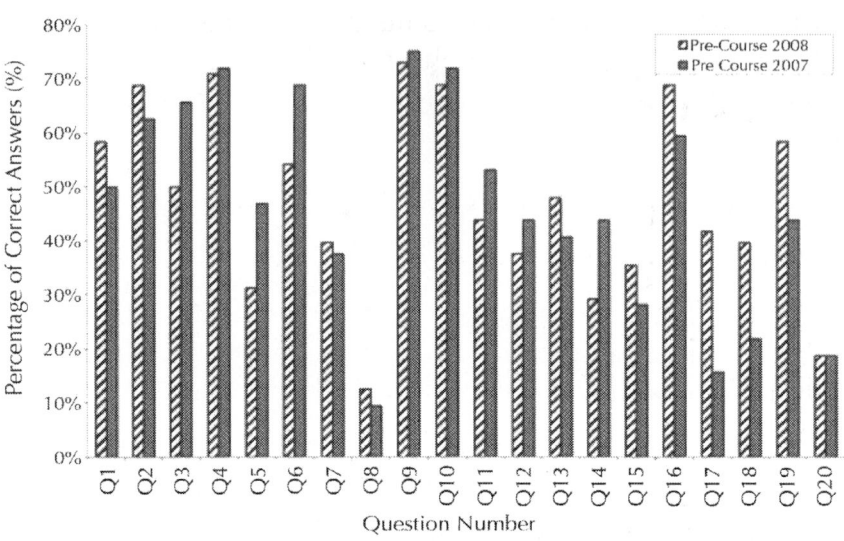

Figure 3: Comparison of pre-course concept inventory results 2007–2008.

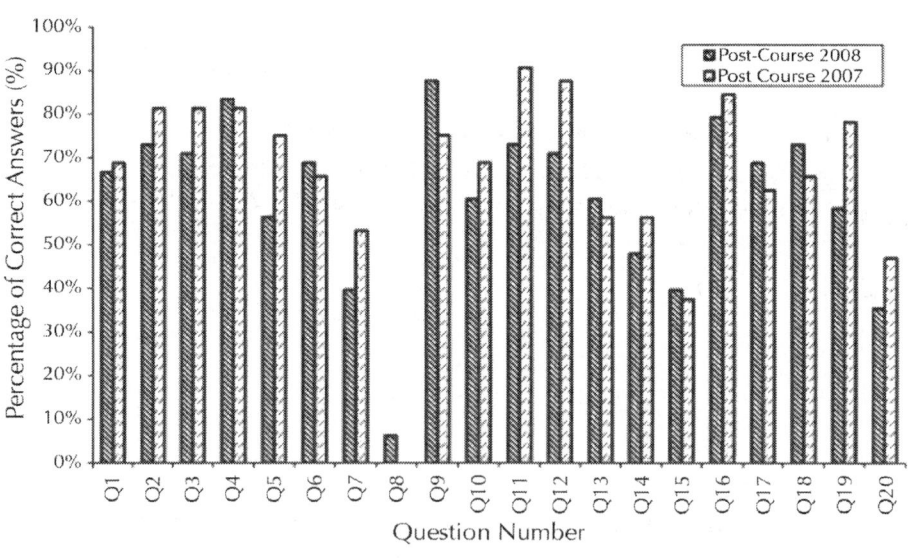

Figure 4: Comparison of post course concept inventory results 2007–2008.

Fig. 3 shows that comparative results were seen for questions 4, 9, 10 and 20. Whilst the 2008 cohort achieved higher in questions 16, 17, 18 and 19 with the 2007 cohort performing better in questions 5, 6 and 14.

It is interesting to note in Fig. 4 that the 2008 cohort achieved superior results in questions 8 and 9, whilst the 2007 cohort achieved better in 5, 11, 12 and 19. This is interesting as the 2007 cohort were superior in a greater number of questions post course, whilst were superior in fewer questions pre course. One of these questions, 19, cohort 2007 were superior post course, whilst cohort 2008 were superior pre course.

STATISTICAL ANALYSIS OF INVENTORY RESULTS

Introduction

A practical consideration that arises when designing a new inventory is the question of how effective the instrument is at garnering valid information that reflects students real conceptual understanding of the ideas tested. This too is the case for the CEFCI where it was determined statistical analysis of the results would acquire data to facilitate an improved understanding of the effectiveness of the CEFCI as a learning tool. Statistical analyses considered include:

Point-biserial correlation. The point-biserial coefficient has been discussed in the literature by Midkiff et al. (2001) and Martin et al. (2004). The point-biserial correlation is a measure indicating the association of individual responses on each question and the students' total result on the CEFCI (Martin et al., 2004). A high positive point-biserial value indicates a strong correlation between those that chose the considered answer and those that were successful on the complete inventory.

Conversely, a low or negative value suggests a strong correlation between those selecting the considered answer and those lacking success in the CEFCI.

t-Test. As stated by Hernon (1991, p. 119) 'The t-test... determines whether or not there is a significant difference between two sample means.' The t-test provides statistical validation of whether the sample mean of the post-test results is larger than the sample mean of the pre-test results. A t-test comparison of the pre and post-test results from 2007 to 2008 provides quantitative evidence for or against the conclusion of cross year improvement.

Point Biserial Analyses

The suggested range for acceptable point-biserial values varies widely in the literature. It has been suggested that point-biserial values for the correct answer on a question ranges from +0.15 and over up to +0.40 and above (Reckase and McKinley, 1991 and Hills, 1976). Selection of an acceptable point-biserialvalue was constrained by the knowledge that the questions in the inventory were constructed by a non-expert. However an acceptable value of +0.15 was chosen. Whilst this value was chosen somewhat arbitrarily with some cursory support from the literature, the considered analysis of all results will ensure that discrepancies, even where the base value of +0.15 is met, will be considered. Where the questions do not meet this prescribed benchmark they have been re-examined. Similarly if the point-biserial value for the distracters do not take a negative value or fail to be selected, they have been reconsidered.

Analysis of the point-biserial value results conveniently presented in Table 1 indicates that all questions in the inventory have point-biserial values within the acceptable range. Apart from questions 3, 8, 10 and 11, all questions have point-biserial values that are considerably greater than the +0.15 benchmark, indicating the strength of the inventory.

Table 1: Point biserial and p-value analysis of inventory results

Item		A	B	C	D	E
1	r_{pbi}	−0.3653	−0.3880	0.3812	−0.2175	−0.00135
2	r_{pbi}	−0.3059	−0.3713	0.2994	−0.1719	−0.1305
3	r_{pbi}	0.1510	−0.5305	−0.3076	Not Chosen	0.0024
4	r_{pbi}	−0.4205	Not Chosen	0.4278	−0.1351	−0.3628
5	r_{pbi}	−0.2557	−0.4362	−0.1614	0.3191	−0.02575
6	r_{pbi}	−0.2343	−0.4267	0.3064	−0.0562	−0.1527
7	r_{pbi}	−0.0562	0.2719	−0.2691	−0.3578	−0.1752
8	r_{pbi}	Not Chosen	−0.2908	0.0801	0.1643	−0.3250
9	r_{pbi}	−0.3279	−0.1912	0.2878	Not Chosen	−0.2709
10	r_{pbi}	−0.3719	0.1543	−0.0562	−0.1852	−0.0353
11	r_{pbi}	0.01960	−0.3502	−0.2162	0.1572	−0.0900
12	r_{pbi}	−0.2946	0.4641	−0.1894	−0.4691	−0.1623
13	r_{pbi}	−0.1930	−0.2830	−0.4275	0.4534	−0.0415
14	r_{pbi}	−0.3753	−0.1871	−0.1894	−0.3810	0.5514
15	r_{pbi}	0.3516	−0.1442	−0.1796	−0.1665	−0.4307
16	r_{pbi}	−0.4103	−0.2542	0.4280	−0.1556	−0.1351
17	r_{pbi}	0.3582	−0.2794	−0.0315	−0.1641	−0.3280
18	r_{pbi}	−0.1825	0.4002	−0.2973	0.3325	0.0024
19	r_{pbi}	−0.2029	−0.2295	−0.2231	−0.1641	0.3656
20	r_{pbi}	−0.0763	−0.2267	−0.1670	−0.0285	0.2381

Bold results indicate the correct answer for that question.

The less satisfactory performance of questions 3, 8, 10 and 11 may be concurrent with the poor performance of some distracters in these questions, as given further consideration below.

Construction and effective use of distracters in a concept inventory paper can be critical to the overall reliability of the inventory. Distracters that students can see are obviously wrong are not effective (Varma, S n.d.). In the CEFCI non-effective distracters are present in questions 3, 4 and 9 where one possible answer was not chosen. Furthermore, it was seen for distracters in questions 8, 11 and 18 that at least one option had a positive point-biserial value. Detail of such a question and the development approach taken can be readily summarised as follows:

CEFCI Question 18

Carbon monoxide at a temperature of 88 °C enters a pipe network at a flow rate of 32 kg/min. The carbon monoxide flows 150 m along a length of pipe before being mixed with a stream of hydrogen. The hydrogen entered the pipe network at a temperature of 65 °C with a flow rate of 4 kg/min a distance 80 m from the mixer. Within the pipe network the temperature and pressure of the gases varies significantly. What is the flow rate of the mixed gas stream at a point 200 m from the mixer?

- Less than 36 kg/min
- 36 kg/min
- Greater than 36 kg/min
- Impossible to say as it depends on the relative diameters of the pipes
- Impossible to say as it depends on the nature of the gas

Students could identify that the mass leaving, in this instance, would be independent of pipe diameter and was not dependent on the nature of the gas. The other options acted as sufficient distracters. One of the non-effective distracters was replaced with an 'I do know option'. The other was replaced with 'Impossible to say as the flow rate will be dependent on the temperature'. It is believed that this option will act as a better distracter for those that do not adequately grasp the conservation of mass principle. It has been seen in the past that students without a comprehensive understanding of the conservation principle are likely to make incorrect assertions regarding the effect of temperature on mass flow rate.

The general principle adopted, to rectify similar issues to those in question 18, was to replace the unchosen answers and the option with the lowest point-biserial value in each question with an 'I don't know' option. This served two purposes; firstly it removed non-effective distracters, ideally improving the effectiveness of these questions. This also served to improve result reliability by reducing the chance of arbitrary guessing of answers. This should also have the effect of improving the point-biserial value of questions 3, 8, 10 and 11 by preventing arbitrary guessing, whilst improved distracters will prevent students with a lack of fundamental understanding been driven unintentionally towards the correct answer.

t-Test Analysis

Quantitative designation of improvement pre and post course for a designated cohort and between cohorts is possible with application of a t-test analysis.

The t-test analysis was produced using Microsoft Excel software and facilitated by a paper composed byMorgan and Deming (2006).

All the t-tests conducted are one sided two sample t-tests. This means that the hypothesis will be:

Null hypothesis. Mean sample 2 $\leq$ mean sample 1

Alternative hypothesis. Mean sample 2 > mean sample 1

The results, in Table 2 and Table 3, show that the value of t_{calc} > t_{crit}. Hence the null hypothesis should be rejected. Furthermore probability of obtaining a value of t that is greater than or equal to t_{calc} by chance when the null hypothesis is true is the p-value. It is observed that the calculated p-value is less than the confidence interval; therefore we can safely reject the null hypothesis at the 95% confidence level (Morgan and Deming, 2006). Therefore the mean of the post course results is judged to be statistically greater than the mean of the pre-course results, at the 95% confidence level, for both the 2007 and 2008 results. This provides quantitative evidence of improved fundamental understanding achieved by attending the Chemical Process Principles II course, as was envisaged.

Table 2: Pre-course and post-course t-test results 2007

Pre–post 2007	Statistics
p	1.298×10^{-5}
t_{calc}	4.534
t_{crit}	1.998

Table 3: Pre-course and post-course t-test results 2008

Pre–post 2008	Statistics
p	5.26×10^{-4}
t_{calc}	3.591
t_{crit}	1.986

The results, in Table 4 and Table 5 are done for a t-test assuming unequal variances. The results in Table 4 show that the value of t_{crit} > t_{calc}. Hence the null hypothesis should be accepted. Furthermore probability of obtaining a value of t that is greater than or equal to t_{calc} by chance when the null hypothesis is true is the p-value. It is observed that the calculated p-value is greater than the confidence interval (Morgan and Deming, 2006). Therefore it is judged that there is no significant difference between the 2007 and 2008 cohorts pre-course results. A similar conclusion can be drawn for the mean of the post course results from 2007 to 2008 as t_{crit} > t_{calc}. Hence it is concluded that the null hypothesis is to be accepted and hence there is no statistical difference between the 2007 and 2008 post-course results.

Table 4: Pre-course 2007 and pre-course 2008 t-test results

Pre 2007–pre 2008	Statistics
p	0.00305
t_{calc}	0.718
t_{crit}	1.997

Table 5: Post-course 2007 and post-course 2008 t-test results

Post 2007–post 2008	Statistics
p	0.274
t_{calc}	1.102
t_{crit}	1.990

The results suggest that there was no statistical difference between the pre-course results from the 2007 and 2008 cohorts. This is expected as both classes completed the same first fundamental Chemical Engineering course in their first year. This suggests that the retention rates of concepts for each year are similar.

The post course statistical analysis suggests that the 2007 and 2008 groups had no statistically significant difference between their post-course results. This suggests that both groups gained similar advantages in knowledge from the Chemical Process Principles course across the two years. This is unsurprising, as teaching methodologies across the two years differed little. However it is desired to use the results presented in Table 1, Table 2, Table 3 and Table 4, to identify key areas, where fundamental knowledge is lacking, in which to implement modified teaching methods to improve the outcomes in these areas.

Overall however and analysis of the mean values indicates that the 2007 group made an overall 42% improvement in their mean mark, whilst the 2008 group improved by 29%. These improvements are both deemed to be significant improvements.

COMPARISON WITH PREVIOUS WORK

This work on concept inventories has been completed in parallel to work completed by Dr. David Shallcross, with his results being from a different cohort of students (Shallcross, 2010). This work is in conjunction with a further concept inventory for first year students, in which a collaborative paper has been published by us and Dr. Shallcross, a paper primarily written by Mr Mark Davis, based on research from Dr. Ngothai & Shallcross (Ngothai et al., 2009).

The inventories used are primarily the same, with some minor changes in structure and wording of some questions. The inventory can be viewed in Appendix A.

There are some distinct similarities and differences between the work completed by ourselves and that completed by Dr. Shallcross. The research undertaken was conducted in a comparative manner with both studies been completed with students on a voluntary basis, in a class setting. However, a primary difference was how the data was utilised. Dr Shallcross used graphics and tables to outline the performance of students, with a greater focus on how students performed from pre to post test. In comparison, this work focussed more on the effectiveness of the inventory itself by using a number of years of data, and statistical tools to determine the effectiveness of questions and answers. However, it is interesting to note that both papers identify inadequacies in similar questions, e.g. Questions 8, in both inventories, are the same question, and were found to be operating with one distracter not effective.

Whilst there are distinct similarities between the works, both papers provide different insights into the use of concept inventories, whilst the use of work from different groups of students over a longer period, in this paper, helps cement some of the findings whilst adding to the understanding provided by the work of Dr. Shallcross.

CONCLUSIONS

The results from the analysis of the concept inventory have demonstrated that the careful consideration of the construction of the inventory proved beneficial. The statistical results underscore the ability of the concept inventory to provide practical data to enable knowledge areas where improved teaching methods will provide benefits, to be identified.

The point-biserial results demonstrated the reliability of the correct answers and the distracters. The results did underscore the questions in which there where distracters that were not operating effectively. Statistical analysis using a t-test proved the effectiveness of the Chemical Process Principles II course at improving student understanding. However there was seen to be no improvements between the two years, on a statistical level. However strategies have been adopted to improve the inventory and remove areas that were deemed to be operating unreliably. Furthermore, validation of the inventory will now allow the inventory results to be further utilised to improve areas in which fundamental student knowledge is lacking. This implementation is to be validated with a longitudinal study utilising this instrument.

The CEFCI could be used to assess the knowledge areas of weakness in students and then provide potential intervention strategies to improve the teaching and learning of fundamental concepts in Chemical Engineering. Furthermore, we as the authors see the potential for the development of similar concept inventories for other engineering disciplines. It is believed such inventories could provide the same benefits, of improved understanding of the learning of fundamental concepts in the relevant discipline.

Appendix A. Concept Inventory

The purpose of this multiple-choice test is to help us understand your understanding of basis concepts in this subject. The result of the test will be used to help us to improve the way in which we teach the material in this subject. The results will have no influence on your grades. If you do not know the answer to a question please use your intuition to try to answer the questions anyway.

In the following questions given below, mark ONE correct answer for each question on the attachment one.

- At 300 kPa a substance boils at 70 °C. The substance is observed to boil when the pressure is increases to 600 kPa. At a pressure of 600 kPa, the substance will boil at:

A. The same temperature (i.e., at 70 °C) because for liquids the boiling point is independent of the pressure.
B. A temperature less than 70 °C because for all compounds the boiling point decreases with increasing pressure.
C. A temperature greater than 70 °C because for all compounds the boiling point increases with increasing pressure.
D. 140 °C
E. An unknown temperature. There is insufficient information to answer whether the temperature will be less than, greater than or equal to 70 °C.

• A small beaker containing an acid solution which is 5% H_2SO_4 and 95% water is mixed with a solution in a large beaker which contains some H_2SO_4 and water. The resultant mixture is 11% H_2SO_4 and 89% water. What was the composition of the solution in the larger beaker before mixing occurred?

A. Less than 5% H_2SO_4.
B. Between 5% H_2SO_4 and 11% H_2SO_4.
C. Greater than 11% H_2SO_4.
D. Exactly 5% H_2SO_4.
E. Impossible to say as there is insufficient information.

• Generally if the pressure is held constant the density of a gas does what with increasing temperature?

A. Decreases.
B. Stays the same.
C. Increases.
D. Can't answer as it depends on the type of gas.
E. It may increase or decrease depending on how close it is to its critical point.

• Which ball in the list below has the highest potential energy relative to the ground? Assume that all the balls have the same mass.

A. A stationary ball sitting on a table 80 cm above the ground.
B. A ball moving vertically upwards as it passes through a height 130 cm.
C. A ball moving vertically downwards as it passes through a height of 150 cm.

D. A ball moving horizontally at a height of 120 cm at a speed of 1 m/s.

E. A ball moving vertically upwards as it passes through a height of 149 cm.

- You use breath to blow up an ordinary party balloon at home. Which of the pressures listed below is closest to the absolute pressure inside the balloon?

A. 100 Pa.

B. 1000 Pa.

C. 10000 Pa.

D. 100,000 Pa.

E. 1,000,000 Pa.

- A gas is flowing along through a pipe at a rate 150 litre/min when it enters a heater. When it leaves the heater the temperature of the gas has increased by 50 °C. The pressure is approximately atmospheric. What is the volumetric flow rate of the gas leaving the heater?

A. Less than 150 l/min.

B. 150 l/min.

C. Greater than 150 l/min.

D. Either less than or equal to 150 l/min.

E. Impossible to say as it depends on the nature of the gas.

- Which one of the following statements is false?

A. The boiling point of water at one standard atmosphere is 100 °C.

B. Water in the gas phase cannot exist in a room at temperatures less than 100 °C if the total pressure in the room is one standard atmosphere.

C. Water can only exist at a temperature of 200 °C if the pressure is much greater than one standard atmosphere.

D. The temperature at which water boils increases with increasing pressure.

E. The wisps of 'steam' seen coming off the surface of a hot mug of tea or coffee on a cold day are actually water vapour.

- What is 100 °C expressed in °F?

A. 100 °F.

B. 180 °F.
C. 212 °F.
D. 180 °F or 212 °F.
E. 32 °F.

- Which one of the operations listed below involves the greatest change in energy of a kilogram of H_2O? Assume all operations are conducted at atmospheric pressure.

A. Heating 1 kg of water from 10.0 °C to 30.0 °C.
B. Heating 1 kg of water from 70.0 °C to 90.0 °C.
C. Turning 1 kg of water at 100.0 °C into 1 kg of steam at 100.0 °C.
D. Heating 1 kg of steam from 110.0 °C to 130.0 °C.
E. Heating 1 kg of steam from 170.0 °C to 190.0 °C.

- It is not possible to cool a substance below absolute zero (i.e., 0 K) because:

A. The energy required to do so is just too great.
B. Absolute zero is the lowest temperature that can exist.
C. Temperature below absolute zero are just too dangerous to handle.
D. Any substance will just break its individual atoms as the temperature is cooled passed absolute zero.
E. We just don't know how to do it.

Questions 11 and 12 refer to the diagram below. The flow rate of stream 3 is greater than the flow rate of stream 1, and the composition of streams 2 and 5 are identical.

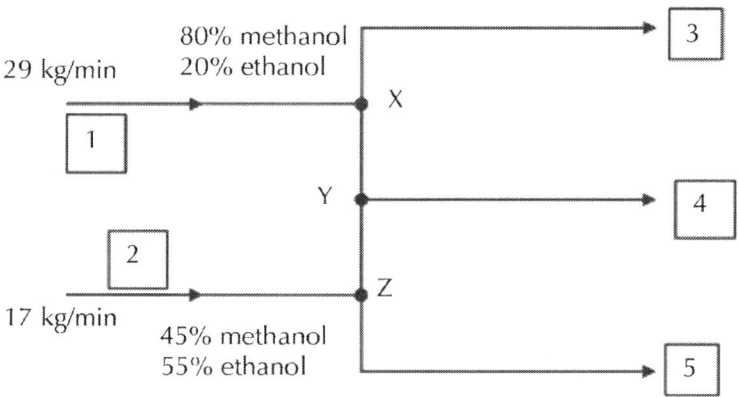

- What are the directions of flow between points X and Y, and between points Y and Z?
A. X to Y and Y to Z.
B. X to Y and Z to Y.
C. Y to X and to Y to Z.
D. Y to X and to Z to Y.
E. Insufficient information is supplied to answer the question.
1) Which one of the following statements is correct?
A. The composition of stream 3 is the same as stream 1.
B. Stream 3 is between 45% and 80% methanol.
C. The composition of stream 4 is the same as stream 1.
D. The flow rate of stream 4 is between 17 and 29 kg/min.
E. The composition of stream 4 is 75% ethanol.
- Refrigerant 123 or 2-2-dichloro-1-1-1-trifluoroethane has the formula CCl_2HCF_3. Which one of the following statements is false?

A. 1 mol of CCl_2HCF_3 contains 4 different elements.
B. 1 mol of CCl_2HCF_3 contains 8 mol of atoms.
C. 1 mol of CCl_2HCF_3 contains 1 mol of CCl_2H.
D. 1 mol of CCl_2HCF_3 contains 3 mol of F_3.
E. 1 mol of CCl_2HCF_3 contains 2 mol of Cl.
- A gaseous mixture of hydrogen and helium enters a non-insulated pipe at a rate of 2.00 m³/h. A distance 300 m down the pipe, the pipe branches. Within the pipe network the

temperature and pressure of the gas vary considerably. If 0.90 m³/h of the gas leaves along branch A, what is the flow rate of the gas leaving branch B?

A. 1.10 m³/h.
B. Less than 1.10 m³/h.
C. Greater than 1.10 m³/h.
D. Less than or equal to 1.10 m³/h.
E. Impossible to say as there is insufficient information.

- The energy of a material increases when its velocity, height or temperature is increased. Suppose you have 1 kg of water. Which of the following operations listed below will increase the water's energy by the greatest amount?

A. Increasing its temperature by 2 °C from 10 °C to 12 °C.
B. Increasing its velocity by 2 m/s from 10 m/s to 12 m/s.
C. Increasing its velocity by 1 m/s from 20 m/s to 21 m/s.
D. Increasing its height above the ground by 2 m from 10 m to 12 m.
E. Increasing its height above the ground by 1 m from 20 m to 21 m.

- Two chemicals, A and B, react in the gas phase to produce a gas C, according to the reaction:

$A(g) + B(g)\ \ 3C(g)$

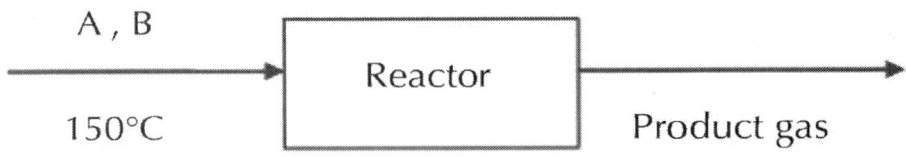

A , B

Reactor

150°C   Product gas

The reaction is exothermic meaning that heat is evolved within the reaction. The reaction occurs with a perfectly insulated reactor so that no heat may enter or leave the reactor through the reactor walls. If reactants A and B enter the reactor at a temperature of 150 °C, at what temperature will the product gases leave the reactor?

A. Less than 150 °C.
B. 150 °C.

C. Greater than 150 °C.
D. Impossible to say as the temperature depends on the speed at which A, B and C enter and leave the reactor.
E. Impossible to say as we are not told if all the A and B are consumed.

• Consider the process shown below.

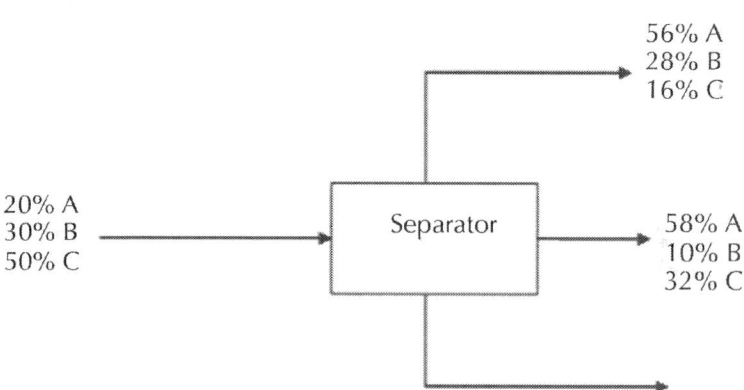

A stream consisting of three components, A, B and C, enters a separator. The composition of the stream is 20% A, 30% B and 50% C. The separator produces three streams of differing compositions, and of differing flow rates. The composition of one of the three streams leaving the separator is 56% A, 28% B and 16% C. The composition of another of the three streams is 58% A, 10% B and 32% C. Which of the compositions listed below is a possible composition of the third product stream?

A. 2% A, 35% B and 63% C.
B. 10% A, 25% B and 65% C.
C. 22% A, 25% B and 53% C.
D. 28% A, 12% B and 60% C.
E. 30% A, 45% B and 25% C.

• Carbon monoxide at a temperature of 88 °C enters a pipe network at a flow rate of 32 kg/min. The carbon monoxide flows 150 m along a length of pipe before being mixed with a stream of hydrogen. The hydrogen entered the pipe network at a temperature of 65 °C with a flow rate of 4 kg/

min a distance 80 m from the mixer. Within the pipe network the temperature and pressure of the gases varies significantly. What is the flow rate of the mixed gas stream at a point 200 m from the mixer?

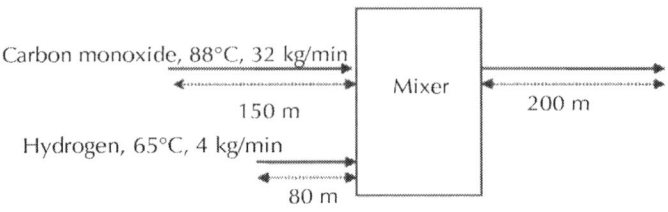

Carbon monoxide, 88°C, 32 kg/min

150 m

Mixer

200 m

Hydrogen, 65°C, 4 kg/min

80 m

A. Less than 36 kg/min.
B. 36 kg/min.
C. Greater than 36 kg/min.
D. Impossible to say as it depends on the relative diameters of the pipes.
E. Impossible to say as it depends on the nature of the gas.

• Which ball in the list below has the lowest kinetic energy relative to a platform on top of a high tower? Assume that all the balls have the same mass and density.

A. A ball moving vertically upwards at a speed of 1.0 m/s as it passes through a height 2 m below the top of the tower.
B. A ball moving vertically downwards at a speed of 0.8 m/s as it passes through a height 3 m above the top of the tower.
C. A ball moving vertically downwards at a speed of 1.2 m/s as it passes through a height 2.5 m below the top of the tower.
D. A ball moving horizontally at a speed of 0.4 m/s and a height of 3 m above the top of the tower.
E. A stationary ball sitting on the platform at the top of the tower.

• An insulated vessel contains 200 ml of water at 60 °C. Into the vessel 200 ml of an acid at 20 °C are poured. The contents of the vessel are then well mixed. If no heat is lost or gained through the walls of the container whilst the liquids are being mixed what will be the temperature of the mixture? Assume no chemical reactions take place.

A. Less than 40 °C.
B. 40 °C.
C. Greater than 40 °C.
D. Less than or equal to 40 °C.
E. Impossible to say as the temperature will depend on the properties of the mixture.

REFERENCES

1. Evans, D., Gray, G., Krause, S., Martin, J., Midkiff, C., Notaros, B., Pavelich, M., Rancour, D., Reed-Rhoads, T., Steif, P., Streveler, R., Wage, K., 2003. Progress on concept inventory assessment tools. In: Proceedings of 33rd Frontiers in Education Conference, ASEE/IEEE, Boulder, CO, pp. T4G1–8.

2. Felder, R., Rousseau, R., 2005. Elementary Principles of Chemical Processes, 3rd ed. John Wiley & Sons, United States of America.

3. Hernon, P., 1991. Statistics: A component of the research process. Ablex Publishing Corp, Norwood, NJ.

4. Hills, J., 1976. Measurement and Evaluation in the Classroom. Merrill, Columbus, OH.

5. Jacobi, A., Martin, J., Mitchell, J., Newell, T., 2003. A concept inventory for heat transfer. In: Proceedings of 33rd Frontiers in Education Conference, ASEE/IEEE, Boulder, CO, pp. T3D12–16.

6. Krause, S., Tasooji, A., Griffin, R., 2004. Origins of misconceptions in a materials concept inventory from student focus groups. In: Proceedings, ASEE Annual Conference.

7. Nelson, W.H., 1997. The Gimli Glider. Soaring Magazine.

8. Martin, J., Mitchell, J., Newell, T., 2004. Work in Progress: analysis of the reliability of the fluid mechanics concept inventory. In: Proceedings of 34th Frontiers in Education Conference, ASEE/IEEE, Savannah, GA, pp. FIF3–4.

9.   Midkiff, K., Litzinger, T., Evans, D., 2001. Development of engineering thermodynamics concept inventory instruments. In: Proceedings of the 31st Frontiers in Education Conference, ASEE/ISEE, Reno, NV, pp. F2A–3.

10. Morgan, S., Deming, S 2006 Guide to Microsoft Excel or calculations, statistics and plotting data, Accessed Online 20th January 2006, http://www.chem.sc.edu/faculty/morgan/resources/excel.

11. Ngothai, Y., Shallcross, D., Davis, M., 2009. Concept inventory for fundamentals of material and energy balances. In: World Congress of Chemical Engineering, Montreal, Canada.

12. Olds, B., Streveler, R., Miller, R., Nelson, M., 2004. Preliminary results from the development of a concept inventory in thermal and transport science. In: Proceedings of the 2004 ASEE Conference, Washington DC, session 3230.

13. Reckase, M., McKinley, R., 1991. The discriminating power of items that measure more than one dimension. Applied Psychological Measurement 15 (4), 361–373.

14. Shallcross, D., 2010. A concept inventory for material and energy balances. Education for Chemical Engineers 5 (1), 1–12.

15. Streveler, R., Geist, M., Ammerman, R., Sulzbach, C., Miller, R., Olds, B., Nelson, M., 2006. Identifying and investigating difficult concepts in engineering mechanics and electrical circuits. In: Paper Presented at the Annual Meeting of the American Society for Engineering Education, Chicago, IL.

16. Varma, S n.d., Preliminary item statistics using point-biserial correlation and p-values, Accessed Online 2nd February 2009, http://www.eddata.com/resources/publications/EDS Point Biserial.pdf.

Chapter 4

Fundamentals of the Knowledge about Chemical Additives Present in Rubber Gloves

Hegles Rosa de Oliveira[I] and Alice de Oliveira de Avelar Alchorne[II]

[I]M.Sc., Escola Paulista de Medicina (São Paulo School of Medicine) - Universidade Federal de São Paulo (Federal University of São Paulo - UNIFESP) - (SP); Dermatologist, Instituto Medico Salette (Salette Medical Institute) - São Paulo - (SP), Brazil

[II]Faculty Member - Professor of Dermatology, Faculdade de Medicina da Universidade Nove de Julho (School of Medicine, Nove de Julho University - UNINOVE) - São Paulo (SP), Brazil

ABSTRACT

Backgrounds

One of the most frequent causes of allergic contact dermatitis of occupational origin are rubber additives, which are present in Personal Protective Equipment (PPE). The most allergenic additives of natural and synthetic gloves are thiurams, carbamates and mercapto group. OBJECTIVE: To investigate the state of knowledge about the chemical additives used in the manufacture of synthetic rubber gloves.

Methods

This was a qualitative research study in which professionals working in the manufacture, research, prescription and commercialization of gloves answered an open questionnaire.

Results

30 individuals were interviewed: 4 researchers in occupational medicine, 5 occupational physicians, 2 occupational safety technicians, a rubber workers' union physician, an occupational safety engineer, a pro duction engineer of rubber gloves, 4 importers of gloves, a manufacturer of gloves, 3 businessmen who sell PPE, 3 salesclerks working in stores that sell PPE, 2 businessmen who own stores that sell products for allergic indivi duals, and 3 dermatologists.

Conclusion

Knowledge of the chemical composition of rubber gloves is scant. The labeling of gloves, with the description of their chemical composition, would facilitate choosing the best type of glove for each person. This low-cost action to businesses would be a gain from the standpoint of public health, with huge repercussions for users of rubber gloves.

INTRODUCTION

Occupational dermatoses represent a significant portion of occupational diseases worldwide. We have observed that its social importance has increased due to improvement in detection. Etiologic diagnosis, in association with the activity performed by the individual, partly depends on the knowledge that the patients have about the products they use and their awareness about their own health condition and risks. It also depends on the knowledge that the physician who is providing assistance has.[1]

Occupational dermatoses represent from 13% to 34% of occupational diseases worldwide. Contact dermatitis represents from 4% to 7% of all dermatologic visits and of these, 50% are occupational contact dermatitis. The family and individual socioeconomic status of workers with allergic contact dermatitis is usually lowered. Individuals begin to experience financial difficulties that affect their health and psychological state. Countries such as the United States spend around 3 billion dollars per year on occupational skin diseases. Studies show that in the United States, Canada, Australia and other countries in Europe 5% to 10% of the population are allergic to rubber additives, and it has been observed that the increased use of personal protective equipment (PPE) is directly related to a higher incidence of allergic contact dermatitis. [2, 3]

In Brazil there are no reliable statistical data on occupational diseases. Some of the reasons for this are lack of access to health services, self-medication and high levels of informal work. To further complicate things, there is the difficulty in establishing occupational etiological diagnosis and lack of notification. However, it is usually possible to identify which chemical substances cause occupational dermatosis. The ones that most often cause allergies are the additives used in the manufacture of rubber. [4, 5, 6] This study aims to explore the social relevance of allergic contact dermatitis triggered by additives present in synthetic rubber gloves. Our objective was to investigate the state of knowledge about these additives by professionals who work directly with synthetic rubber gloves, either in research, manufacture, sales or recommendation of its use as PPE. This topic is important because skin diseases are a cause of workplace absenteeism. Moreover, if workers become ill, they can be permanently professionally disabled. [7, 8, 9]

MATERIAL AND METHODS

The objective of this study, which was "the understanding of the actual state of knowledge of each interviewee about the chemical additives used in the manufacture of synthetic rubber gloves," was achieved by means of qualitative research using questionnaires applied through interviews. Data was collected via personal contact with each interviewee through the application of a questionnaire. We interviewed 30 individuals with potential knowledge about synthetic rubber gloves and their chemical composition: 4 researchers in occupational medicine, 5 occupational physicians, 2 occupational safety technicians, a rubber workers' union physician, an occupational safety engineer, a production engineer of rubber gloves, 4 importers of gloves, a manufacturer of gloves, 3 merchants who sell PPE, 3 salesclerks working in stores that sell PPE, 2 owners of stores that sell products for allergic individuals, and 3 dermatologists.

In addition to the interviews, we contacted six synthetic rubber gloves manufacturing factories to request access to their production sector.

RESULTS

The types of synthetic rubber gloves described and known by respondents were nitrile, neoprene, butyl, Viton®and Silver Shield.

® 10 Protective glove testing for the issuance of the CA (Certificate of Approval) in Brazil include mechanical resistance tests and, in the case of sterile surgical gloves, microbiological assays, based on information from a researcher who works in the area of technological development.

Knowledge of most respondents is limited, as they do not know which chemical additives are present in synthetic rubber gloves. Physicians, occupational safety technicians, occupational safety engineers, production engineers working for a glove manufacturing industry and merchants, in general, had little practical and theoretical knowledge on the subject. This reinforces the findings of the literature.[11] only the first two respondents who carry out research in the field of occupational medicine had extensive experience on the topic and could enrich the research. They reported that in recent years there has been a huge increase in the use of PPE; however, despite the fact that Brazilian legislation regulates the use of protective gloves, often what weighs more when a company chooses the type of glove they will buy is the presence of CA (Certificate of Approval), which is required by law for PPE and to determine costs.

Based on an informal conversation with a researcher, the authors of the present study were told that there had been no major changes in the chemical additives used in the manufacture of gloves in recent years. We were also informed that thiourea is an old allergen, but the increase in publications classifying thiourea as a major allergen, which is present in neoprene rubber, indicates that this rubber is being more widely used today.

The protective gloves that dermatologists are more familiar with are surgical latex gloves and gloves for household use.

Occupational physicians are trained to target prevention. They have a little more knowledge about the universe of protective gloves, but it is more theoretical. The gloves they recommend in case of allergy to natural rubber are PVC gloves (polyvinyl chloride). These physicians can have more access to workers. They state that many workers have reported difficulty making use of gloves continuously for many hours, complaining that the glove heats up and their hands sweat. If the skin is injured, it is difficult to wear gloves and manual dexterity is impaired; sometimes, the use of gloves reduces production.

The occupational physician who works in the area of labor issues concerning workers in the rubber industry states that these workers have other diseases, in addition to contact dermatitis caused by allergy to rubber additives.

Occupational physicians stated that the use of gloves for hand protection not always complies with pre-established scientific criteria. They agree that good quality, carefully chosen gloves will sell faster. The correct choice of gloves and the chemical substances used to manufacture them, the perfect finish of the final product, as well as knowledge and education of the worker who will use PPE are important factors for adhesion to the use of protective gloves and PPE.[12, 13]

Salesclerks who work in stores that sell PPE and were interviewed by us did not receive training and information from their employer. The knowledge they had was passed on orally by more experienced coworkers. Their knowledge is informal and they only provide general information to the buyer.

The storeowner who sells products for allergy sufferers stated that she sold cotton gloves to be worn under rubber gloves when buyers reported that they were allergic to rubber. The researcher of this project explained that cotton does not offer protection, since the allergen could easily break the barrier and reach the skin. She advised it is best to use delicate vinyl gloves.

All the respondents, in a conversation that went beyond the scope of the survey, agreed that if PPE had a label indicating the chemical composition of the product, similar to what happens with food and drugs, it would be easier to recommend the best type of protective gloves to those sensitized to any chemical additives of rubber. The labeling of gloves is an act of citizenship that would not increase manufacturing costs. It would add value to the product, characterizing socially responsible marketing. The author interviewed a production engineer working in the glove manufacturing industry who is also a member of an international organization that studies rubber gloves. She informed us that even abroad the identification of the chemical composition of rubber gloves is an issue that generates a lot of debate and is full of controversies due to fear of competition.

One of the respondents said that access to information is a right guaranteed by the consumer code, but this aspect of democracy and respect for citizens has not yet been considered in the manufacturing of protective gloves. Access to information, especially by manufacturers of synthetic rubber gloves, was disappointing. No company contacted provided us with manufacturing details.

DISCUSSION

The rubber present in gloves can cause different types of skin reactions. The use of gloves by susceptible individuals can trigger irritative contact dermatitis, late allergic contact dermatitis caused by vulcanizing agents or an immediate reaction. Of the five types of synthetic rubber gloves, nitrile and neoprene gloves are more easily found, whereas butyl, Viton® and Silver Shield® are harder to find because they are less sold and more expensive. Nitrile gloves are the most marketable. Silver Shield ® and Viton ® gloves are special gloves which have high resistance to various chemical substances, are used to protect against "hazardous" chemical agents or recommended when the professional is dealing with an accident with chemical substances and does not know the specific type of risk involved. Silver Shield® is a thinner, disposable glove that can be worn over another glove. [14]

The country that has conducted more studies on allergens is Japan, where they use chromatography to identify the chemicals in gloves. Sometimes they combine the chromatography and patch tests to investigate the possible causes of allergic contact dermatitis. [15]

Some researchers advise that when an individual is suspected of having contact dermatitis caused by rubber additives, but the patch test with a standard battery of patches is negative, thiourea may be the suspected allergen.[16, 17] A study of this population showed that this group presents a higher incidence of stomach and upper aerodigestive tract cancers when compared with the unexposed population.[18]

The choice of gloves is a complex act, assuming that there are several brands of rubber gloves available in the market, each with different chemical and physical characteristics. No glove is resistant to all types of substances. [19]

It is true that prolonged use of gloves has advantages and disadvantages, but in 1990 the author stated that this must be taken into consideration when choosing the type of glove that the worker will wear. [20] Studies show that dermatologists have little knowledge about the different types of gloves and about their selection criteria.[21]

It is important to consider that the best information on the chemical composition of protective gloves will always be given or should be provided by the manufacturer. A list of chemical substances may be provided by the toxicological records of the product.[22]

Today it is clear that the use of PPE has increased due to safety and health policies in the workplace, which are slowly advancing both with regard to entrepreneurs and workers.

All the respondents, in a conversation that went beyond the scope of the survey, agreed that if PPE had a label indicating the chemical composition of the product, similar to what happens with food and drugs, it would be easier to recommend the best type of gloves to those sensitized to any chemical additives of rubber. The labeling of gloves is an act of citizenship that would not increase manufacturing costs. It would add value to the product, characterizing socially responsible marketing. The benefits are enormous, ensuring that manufacturers sell only good quality products with labels specifying the chemical components used in the manufacturing process of the product and its indicated use for each activity, similar to a drug that comes with specifications about its chemical composition, adverse reactions, and indications of use so that the marketing is socially responsible and sales can increase. This will also guarantee that the product is approved by quality control agencies.

Employers can be glad to know that their employees are being offered a quality product which is comfortable, thus minimizing skin lesions and promoting the use of PPE for the hands, with an increase in workers' adherence to this equipment. In the end, this may result in lower absenteeism, increased productivity and healthier and more satisfied workers. [23] It is encouraging to see that some companies are beginning to listen to their employees. This is a positive thing that interferes with the entire chain. This change of attitude will benefit everybody - from the supplier who sold the product to those who will wear it and have better working conditions, to the employer who did not spend money on PPE only to meet his legal obligations. This might lower absenteeism. A study describing the experience of a company that had a high rate of hand dermatitis due to lack of adherence to the use of PPE proves this. The company invested in education and encouraged workers to choose the type of PPE that they considered most appropriate to perform their activities. The author explains that before the workers chose the best type of PPE, the work safety team had conducted previous research and selected the most appropriate PPE for the activity to be performed. Inviting the worker to take part in the selection of PPE caused an increase in adherence to the use of this equipment. Subsequently, the number of hand dermatoses decreased and the workers felt they were heard and co-responsible for the gains.

The Brazilian labor law regulates that companies should increase their investments to implement collective protection measures, that is, substitution of allergenic substances, change of machinery, etc. In the short term, while these collective measures are not implemented, the use of PPE is necessary. The rule refers to changes in the workplace, but what we see on a daily basis is the prolonged use of protective equipment, unlike what the law dictates.

The domestic manufacturer or importer should market PPE with technical instructions in the national language, guiding its use, maintenance, restrictions and other references of use. The Ministry of Labor is responsible for supervising and orienting the appropriate use and quality of PPE, among other things.

A study conducted by the Ministry of Labor involving manufacturers and buyers of PPE made a sad reality clear, which only confirms what physicians feel when they recommend the use of the most appropriate PPE for workers and/or their function. The result of this study shows that most manufacturers and companies that buy PPE are only concerned about complying with the legislation. The primary factor guiding the buying decision is the presence of CA, issued by the Ministry of Labor, and then cost. Comfort, lightness and better finish are often not taken into account. It is important to note that small details can interfere with comfort and they can either alleviate or intensify the daily torture of being forced to use PPE. [24]

Since many chemicals are used in the manufacture of rubber gloves, sometimes it is very difficult to identify which allergen is responsible for the allergy. In Brazil, physicians still rely on the patch test for guidance, but the patch test battery is limited because it contains only the most common allergens. With technological advances, every day new products are introduced in the manufacture of rubber. Therefore, even though batteries are constantly reviewed and updated, they will always be deficient. In addition to these factors, we must consider that this test depends on a rigorous training of the person who will administer it, the results take at least 96 hours and it can only be performed when the dermatosis is not in the acute phase. Research is being conducted and in the near future we can expect more technological resources such as chromatography to identify the allergen in gloves and lymphocyte proliferation test, which can identify the antigen using patient serum. [25,26]

The production and consumption of natural and synthetic rubber have increased over the past years.[27] we can infer that contact dermatitis caused by synthetic rubber allergens has also followed this trend. Studies show that the prevalence of allergies to rubber additives and latex proteins has followed an upward curve, particularly after the 80s with the advent of AIDS, when, as preventive measures, rules to minimize the risk of contact with biological secretions were adopted. With this, there was increased use of gloves and allergic contact dermatitis caused by the use of PPE. Early and late allergic responses in health workers are becoming a public health issue. The use of gloves has also increased in the industrial and service sectors. [28] Workers in the civil construction and cleaning areas (including housewives) have more dermatitis caused by the use of protective gloves. Several studies have shown that the main rubber additives involved in triggering delayed hypersensitivity reaction are thiurams, carbamates and mercapto mixes. [29]

Brazilian law says that consumers have the right to know what they are buying or consuming. But this information is not provided by the industry. It is not available for physicians, researchers or the population that buys rubber products. Every chemical substance has a safety data sheet that provides information about the chemical agent, its properties, toxicity and conduct that should be taken in the event of an accident involving such substance. This information should be widely publicized. The entire commercial chain should have access to this information. Even companies should provide these data when requested by health services. Our level of democracy has not yet reached this stage. One way to grant universal access to information would be to label glove packages similarly to what is done with medicine and food.

Attempts to replace allergens with a higher potential for sensitization have been done; an example was the replacement of mercaptobenzothiazole with mercaptobenzimidazole. However, we have observed that the latter has also become an important sensitizer over time. Currently the reduction of the concentration of substances known to be allergens has been suggested. Another factor that hampers the substitution of a product that causes allergy for another that causes less sensitization is high cost. In Denmark, where recommendations of a number of committees have been accepted by European manufacturers, a carbamate (dibutyldithiocarbamate) with less potential to cause sensitization has been used to replace other allergens.[30]

Polysensitization to rubber components - thiuram, mercapto mix, carba mix - is commonly observed in the patch test. These groups of chemical substances are used to manufacture various products used in the workplace and in people's daily lives. This makes it difficult for a person who is allergic to rubber components to abstain completely from contact with these allergens. An individual allergic to any rubber component can be exposed to allergens via medicine, food, pesticides, etc. Today it is almost impossible for a person to spend a day without coming into contact with a product containing rubber substances in its composition. The study of chemical additives used in the manufacturing process of synthetic rubber gloves is important from a practical, economic and political standpoint. Practical because, as individuals become aware of the importance of using gloves to protect their hands, the use of synthetic rubber will increase and, as a consequence, more and more people can become sensitized to rubber additive allergens and to new allergens that may appear as technological changes are implemented in the manufacture of gloves. Economic because sales of gloves will increase; with the availability of good quality gloves the author assumes that the worker will adhere to the use of this PPE, the risks of accidents, hand diseases and absenteeism will decrease, etc., and political because guidance on the use of PPE is part of the Regulatory Norms that control the safety and health of workers in Brazil. In a democratic government citizens have the right to obtain detailed information about what they are consuming and using and which risks a particular product can pose. Therefore, the labeling of gloves with information about the chemical substances used in the manufacture of each type of protective glove is not only politically correct but, first and foremost, an act of citizenship.

CONCLUSIONS

The main additives with the greatest sensitizing potential used in both natural and synthetic rubber gloves are the Thiuram, Mercapto and Carbamate groups. This study points to the scant knowledge about the chemical substances used in the manufacture of rubber gloves. Therefore, better and wider dissemination of the chemical composition of gloves is need. A cheap way to do this would be to standardize a label with the description of the chemical substances present in each type of glove.

REFERENCES

1. Ali AS. Dermatoses Ocupacionais. São Paulo: Fundacentro; 2010.

2. Kadsyn DL, McCarter K, Achen F, Belsito DV. Quality of life in patients with allergic contact dermatitis. J Am Acad Dermatol. 2003; 49:1037-48

3. Hintzenstern JV, Heese A, Kock HV,Peters KP, Hornstein DP.Frequency spetrum and accupational relevance of type IV allergies to rubber chemicals. Contact Dermatitis.1991; 24:244-52.

4. Romaguera C, Grimalt F. Statistical and comparative study of 4600 patients tested in Baecelona (1973-1977). Contact Dermatitis. 1980; 6:309-15.

5. Lammintausta K, Kalimo K. Sensitivity to rubber. Study with rubber mixes and individual rubber chemicals. Derm Beruf Umwelt. 1985; 33:204-8.

6. Grupo Brasileiro de Estudo em Dermatite de Contato (GBEDC) do Departamento Especializado de Alergia da Sociedade Brasileira de Dermatologia. Estudo multicêntrico para elaboração de uma bateria padrão brasileira de teste de contato. A Bras Dermatol. 2000; 75:147-56.

7.   Conde-Salazar L, del-Río E, Guimaraens D, González Domingo A. Type IV allergy to rubber additives: a 10 - years study of 686 cases. J. AM. Acad. Dermatol. 1993; 29:176-80.

8.   Alchorne AOA, Calafiori Jr, Kitamura S, Wakamatsu CT. Alguns aspectos das dermatoses profissionais pelo cimento na construção civil. Rev Bras Saúde Ocup. 1975; 11:1-12.

9.   Macedo MS, de Avelar Alchorne AO, Costa EB, Montesano FT. Contact allergy in male construction workes in São Paulo, Brazil, 2000-2005. Contact Dermatitis 2007; 56:232-4.

10. Iisrp.com [Internet]. IISRP - Internacional Institute of Synthetic Rubber produces - 2007. [Cited 2010 Oct 20]. Available from: www.iisrp.com.

11. Mansdorf SZ. Guideline for selection of gloves for the workplace. Dermatol Clin. 1994; 12:597-600

12. Agner T, Held E. Skin protection programmes. Contact Dermatitis. 2002; 46:253-6.

13. Géralt C, Tripodi D. La prévention des dermatoses professionanelles. La Revue Du Praticun.2002; 52:1446-50.

14. Estlander T, Jolanki R, Kanerva L. Allergic contact dermatitis from rubber and plastic gloves. In: Boman A, Estlander T, Wahlberg JE, Maibach, editors. Protective Gloves for occupational use. Boca Raton: CRC Press; 1994. P.221-9.

15. Kaniwa MA, Isama K, Nakamura A, Kantoh H, Itoh M, Ichikawa M, et al. Identification of causative chemicals of allergic contact dermatitis using a combination of patch testing in patients and chemical analysis. Contact Dermatitis. 1994; 10:20-5.

16. Woo DK, Militello G, and James WD. Neoprene. Dermatitis. 2004; 15:206-9.

17. Patrick E, Robert A. Clinical review: tioureas and allergic contact dermatitis. Cútis. 2001; 68:387-96.

18. Neves H, Moncau JEC, Kaufmann PR, Filho Wünsch V. Mortalidade por câncer em trabalhadores da indústria de borracha de São Paulo. Rev Saúde Pública 2006; 40:271-9.

19. Maso MJ, Goldenberg DJ. Contact dermatoses from disposable glove use: A review. J Am Acad Dermatol. 1990; 23:733-7.

20. Tobler M, Freiburghaus AV. A glove with exceptional protective features minimizes the risks of working with hazardous chemicals. Contact Dermatits. 1992; 26:299-303.

21. Feinman SE. Sensitivity to rubber chemicals. Cutan Ocul Toxicol. 1987; 6:117-53.

22. Estlander T. Jolanki J, Kanerva L. Rubber glove dermatitis: a significant occupational hazard prevention. Curr Probl Dermatol. 1996; 25:170-6.

23. Heron RLJ. Worker education in the primary prevention of occupacional dermatoses. Occup Med. 1977; 47:407-10.

24. Zago JE, Silca JCP. O design pode contribuir para a melhoria dos EPI's. CIPA: Caderno Informativo de Prevenção de Acidentes. 1998; 19:101.

25. Kimber I, Quirke S, Cumberbatch M, Ashby J, Paton D, Aldridge RD, et al. Lymphocyte transformation and thiuram sensitization. Contact Dermatits.1991; 24:164-71.

26. Sanchez APG, Maruta CW, Sato MN, Ribeiro RL, Zomignan CA, Nunes RS, et al. Estudo da proliferação linfocitária em pacientes sensibilizados ao níquel. A Bras Dermatol. 2005; 80:149-58.

27. Tuscott W. They're not just gloves: a guideline on proper use. CDS Rev. 1995; 88:22-9.

28. Heese A, Hintzenstern JV, Peters KP, Kock HV, Hornstein DP. Allergic and irritant reactions to rubber gloves in medical helht services. J Am Acad Dermatol. 1991; 25:81-9.

29. Knudsen BB, Laesen E, Egsgaard H, Menné T. Release of thiurams and carbamates from rubber gloves. Contact Dermatitis. 1993; 28:63-9.

30. Ingber A, Halevy S, Trottner A, Woltfriends S. Mercapo-mix. Contact Dermatitis. 1995; 32:255.

5

Safety Study of an Experimental Apparatus for Extraction with Supercritical CO$_2$

V. B. Soares[I] and G. L. V. Coelho[II]

[I]Laboratory of Separation Process, Department of Chemical Engineering, University Federal Rural do Rio de Janeiro, (UFRRJ), Seropédica, Rio de Janeiro - RJ, Brazil

[II]Laboratory of Separation Process, Department of Chemical Engineering, University Federal Rural do Rio de Janeiro, (UFRRJ), Seropédica, Rio de Janeiro - RJ, Brazil

ABSTRACT

During the process of supercritical CO$_2$ extraction it is necessary to use high pressures in the procedure. The explosion of a pressure vessel can be harmful to people and cause serious damage to the environment. The aim of this study is to investigate the probability of death and injury in a laboratory unit for supercritical fluid extraction in the case of an explosion of the extractor vessel. The procedure is explained via a case study involving fatty acid extraction from vegetable oils with carbon dioxide above its supercritical conditions and under optimum operating conditions. According to the results, more importance should be given to the use of a protective headset because the probability of eardrum injury is superior to the probability of death from lung injury.

INTRODUCTION

Over the last few decades, the word "security" has been substituted by the expression "loss prevention" to include new questions about the safety of the environment, materials and businesses (Hendershot, 2006). This is partially due to a large number of accidents involving pressure vessels in the decade between 1970 and 1980, affecting the image and the profits of organizations. In a chemical manufacturing plant, the sudden release of pressurized material is a negative event, especially when it involves the release of toxic materials or flammables. In these cases, fire, explosions and contamination can also occur (Perry et al., 2008). Obviously, the type of substance, its quantity and the operating conditions will determine the intensity of the impact of an accident. In a supercritical extraction plant, the presence of pressure vessels and other devices is necessary due to the high pressures involved in the process. These devices have a variety of functions in the process such as an extraction vessel, a separator, a heater, a condenser and also as a storage tank for volatiles substances (McHugh and Krukonis, 1986).

The challenge for the equipment manufacturer lies in finding technical solutions which are economically feasible and safe. In recent years, also in Brazil, groups of University laboratories have shown an increasing interest in high pressure technology. On the other hand, generating and containing pressures up to hundreds of bars from gases at room temperature are not routine in Research and Development Labs and require special training. These include an adequate level of shielding, training personnel in operation and preventive maintenance and, more importantly, working within the rated pressures and requiring the use of personal protective equipment such as a respirator, gloves, lab coat, safety glasses etc.

Supercritical Fluid Extraction (SFE) is a process that is growing in importance as an alternative to conventional separation processes. The application of supercritical fluid (SCF) solvents is based on the experimental observation that many gases exhibit enhanced solvating power when compressed to conditions above the critical point. The high selectivity, the speed of the process and the possibility to substitute organic solvents by other less aggressive solvents are some of the characteristics of supercritical technology (Carlès, 2010; Cavalcante et al., 2005; McHugh and Krukonis, 1986; Schneider, 1983; Velasco et al., 2007).

Supercritical carbon dioxide ($SCCO_2$) has been intensively used because of its relatively low critical pressure (73 bar), its low critical temperature (304 K), its non-dangerous character and low cost. Carbon dioxide is considered to be a GRAS solvent with a TLV-TWA value of 5000 ppm (TLV-TWA is the threshold limit value time-weighted average concentration for a normal 8 h workday or 40 h workweek, to which all healthy workers may be repeatedly exposed, day after day, without adverse effect).

As high pressure equipment is installed and operated worldwide, safety rules are internationalized. Regulations for overpressure protection are provided in API RP 520 (American Petroleum Institute), which presents guidance for determining the requirements for the installation, maintenance, and decommissioning of pressure vessels and autoclaves. The pressure relief devices, covered in this recommended practice (RP), are intended to protect unfired pressure vessels and related equipment against overpressure from operating and fire contingencies. The rules for overpressure protection of fired vessels are provided in ASME Section I and ASME B31.1. Also relevant standards for Boiler and Pressure Vessel Code are provided by the Canadian Standards Association (CSA) and the American National Standards Institute/American Society of Mechanical Engineers (ANSI/ASME). In the Federal Republic of Germany, laboratories are required by law to comply with accident-prevention rules of the appropriate professional association and, consequently, are subject to technical regulations published in newsheets of the Pressure Vessel Study Group (AD-Merkblätter, Arbeitsgemeinschaft Druckbehälter). In Brazil, the regulations are provided by the Ministry of Labor and Employment in NR-13 (NR - Norma Regulamentadora).

In this process, the extractor vessel (pressure vessel) is the most important equipment, where the supercritical conditions are established and the extraction occur.

The definitive advantage of supercritical fluids over classical liquid solvents has long been known: the solvent-extract separation is easy and requires very low energy consumption. In fact, this separation is based on the drastic decrease of the solvent power when its specific gravity is decreased: decompression at constant temperature or heating at constant pressure are both used, as explained on Mollier's diagram.

The designer's priority task should be to ensure full safety of operation in the process. Therefore, to achieve this goal, the assessment of the amount of energy stored in the system and the prediction of the scenario of a possible breakdown are essential to protect the environment. Normally the measure of the hazard of a gas working fluid is expressed as the product of the pressure and the volume in the pressure vessel (Radomski and Ros, 1992). Due to the high pressure involved in a supercritical extraction plant, any rupture in the structure of the pressure vessel can cause significant damage in the process area in the case of an explosion.

SCF extraction operations not only require the use of hardware rated for high pressures, but also the employment of multiple pressure-relief mechanisms and safety practices. The most common method of overpressure protection is through the use of safety relief valves and/or rupture disks with discharge directly into the atmosphere and are placed on the SCF pump, pressure vessels and at additional positions containing high pressure. In summary, safety is an important factor while dealing with supercritical extraction systems and the design of such equipment should take all the factors into account.

The aim of this study is to examine the probability of the occurrence of deaths and injuries in a laboratory unit for supercritical extraction in the case of an explosion of the extractor vessel.

The procedure is explained via a case study involving the explosion of an extractor vessel with subsequence formation of overpressure. The optimum conditions (P,T) for the extraction of fatty acids present in vegetable oils using carbon dioxide in supercritical conditions were determined in a previous study, as well as the minimum energy necessary to execute this operation continuously on an industrial scale (Penedo and Coelho, 1997). In order to get the thermodynamics data for a given step of the process, it is particularly useful to have a phase diagram for the solvent used in the extraction. Many diagrams are available for CO_2, but they are not always adequate (Eggers, 1980). Temperature-entropy (T, s) diagrams are particularly appropriate because, in these diagrams, the heat energy supplied and removed in reversible processes is represented by areas. The calculation of the energy released at the moment of the vessel explosion was determined by assuming ideal behavior of the gas and adiabatic expansion (Prugh, 1991). The probability of deaths and injuries was determined using a statistical vulnerability model, the probit method (Lucas et al., 2003; Sierra, 1991).

SAFETY ANALYSIS OF THE USE OF A SUPERCRITICAL EXTRACTION DEVICE

The safety analysis aims to identify, assess and prevent the possible threats involved in the use of a process or device inside or outside an organization. Although the exact nature of the threats and their resulting consequences are difficult to determine, the benefits of a safety analysis are clearly perceived since the prevention of accidents is less costly than to repair the damages afterward (Perry *et al.*, 2008).

Identification of the Threats in a Supercritical Extraction Plant

The analysis of threats is used to identify the sources and types of hazards (Bajpai and Gupta, 2005). Many sources of hazards can be present in a supercritical extraction plant and affect people, the environment, equipment, constructions and business. Some examples of these hazards are a failure in the control of the process, mechanical failure of the equipment, damage to the infrastructure, inappropriate maintenance, electrical damage, undue operation of the safety valve, dust explosion, explosion, boiling liquid expanding vapor explosions (BLEVE), fires (Crowl and Louvar, 2002; Perry *et al.*, 2008; Sklet, 2006), floods, tornadoes, storms, hurricanes, earthquakes (Lucas *et al.*, 2003), terrorist attacks, vandalism, cyber-attacks, sabotage (Bajpai and Gupta, 2005), etc. All of these threats can affect the extraction vessel and compromise its structure. Beside these, excessive elastic deformation, excessive plastic deformation, high temperature, fracture, fatigue and corrosion are also typical problems involved in the use of a pressure vessel that is built following regulation codes.

Consequence Analysis

To estimate the losses and costs caused by an accident is not a trivial task. The damage to equipment, piping, utilities and installations, the effects on the local population, environment, image reputation and loss of profit, and the indemnities among others, must be accounted for in the construction of a safety analysis model (Medina *et al.*, 2009). The prediction of such effects for a specific accident is based on the assessment of vulnerability models already prepared. Among the statistical vulnerability models, the probit function is commonly used to determinate the damages to people and constructions. The probit method is based on experiments carried out with laboratory animals and on studies of the damages caused by previous accidents (Lucas *et al.*, 2003). These vulnerability models allow the determination of the probability of deaths and injuries to people when exposed to fires, explosions and toxic substances release (Sierra, 1991).

In the particular case of explosion of the pressure vessel, the first expected effects on people are eardrum fracture and deaths from lung injury. Secondly, deaths due to body impact, injuries due to body impact, first degree to fatal burns and intoxication are predicted (Lucas *et al.*, 2003; Salzano and Cozzani, 2006; Sierra, 1991). Obviously, these effects depend on the quantity and type of materials, the characteristics of the people involved and the exposure time. The present study will focus only on the estimation of the probability of deaths from lung damage and injury due to eardrum rupture, using Equations (1) and (2), respectively (Lucas *et al.*, 2003; Sierra, 1991).

$$Pr = -77.1 + 6.91 \ln(P) \tag{1}$$

$$Pr = -15.6 + 1.96 \ln(P) \tag{2}$$

Where: Pr is the probit value; P is the maximum overpressure [Pa]. Table 1 can be used to convert the probit value to its respective probability value.

Table 1: Equivalence of the probit values and percentage of population affected

Pr	%	Pr	%	Pr	%	Pr	%	Pr	%
0.00	0	4.19	21	4.8	42	5.33	63	5.99	84
2.67	1	4.23	22	4.82	43	5.36	64	6.04	85
2.95	2	4.26	23	4.85	44	5.39	65	6.08	86
3.12	3	4.29	24	4.87	45	5.41	66	6.13	87
3.25	4	4.33	25	4.9	46	5.44	67	6.18	88
3.35	5	4.36	26	4.92	47	5.47	68	6.23	89

3.45	6	4.39	27	4.95	48	5.5	69	6.28	90
3.52	7	4.42	28	4.97	49	5.52	70	6.34	91
3.59	8	4.45	29	5	50	5.55	71	6.41	92
3.66	9	4.48	30	5.03	51	5.58	72	6.48	93
3.72	10	4.5	31	5.05	52	5.61	73	6.55	94
3.77	11	4.53	32	5.08	53	5.64	74	6.64	95
3.82	12	4.56	33	5.1	54	5.67	75	6.75	96
3.87	13	4.59	34	5.13	55	5.71	76	6.88	97
3.92	14	4.61	35	5.15	56	5.74	77	7.05	98
3.96	15	4.64	36	5.18	57	5.77	78	7.33	99
4.01	16	4.67	37	5.2	58	5.81	79	7.41	99.2
4.05	17	4.69	38	5.23	59	5.84	80	7.51	99.4
4.08	18	4.72	39	5.25	60	5.88	81	7.65	99.6
4.12	19	4.75	40	5.28	61	5.92	82	7.88	99.8
4.16	20	4.77	41	5.31	62	5.95	83	8.09	99.9

RELEASE OF ENERGY IN THE CASE OF AN EXTRACTOR VESSEL EXPLOSION

The intensity of overpressure is closely related to the amount of energy available to generate it. To determine the released energy it is necessary to know the type of equipment, the mechanism of failure, the quantity of material present and the operating conditions. As discussed earlier, boiling liquid expanding vapor explosions or BLEVE can cause the explosion of an extractor vessel in a supercritical extraction plant. This physical phenomenon is characterized by sudden release (Salzano and Cozzani, 2006) and its most common causes are mechanical damage, overfilling, runaway reactions, overheating, vapor space contamination and mechanical failure (Abbasi and Abbasi, 2007).

During the BLEVE process, an instantaneous increase in volume of the substance confined in the vessel occurs due to the expansion of the vapor phase already existing in the vessel. This expanded material releases a high amount of energy converted into overpressure, thermal radiation and missile ejection. Some particular conditions at the moment of the event define the ways in which this energy is distributed. For example, most of the vessels are built with ductile materials and, in this case, 40% of the energy released in a BLEVE is converted into overpressure. The rest of the energy is used to break the equipment, to project the fragments and to generate heat in the environment (Ronza et al., 2007). This rapid deterioration of the tank and tremendous release of energy can propel the tank and the whole equipment to great distances.

There are many ways to estimate the energy released in a BLEVE (Abbasi and Abbasi, 2007; Planas-Cuchi et al., 2004). One possibility is to assume that the vapor behaves as an ideal gas and that the vapor expansion is adiabatic and reversible (Prugh, 1991). In this case, the energy released is obtained from Equation (3).

$$Ev = 10^2 \left(\frac{PV}{\gamma - 1} \right) \left(1 - \left(\frac{Pa}{P} \right)^{\left(\frac{\gamma - 1}{\gamma} \right)} \right)$$

(3)

Where: Ev is the energy released (kJ), Pa is the atmospheric pressure (bar), V is the initial volume of vapor (m³), P is the pressure (bar) in the vessel just before the explosion, γ is the ratio of specific heats.

CASE STUDY

The present case study consists in investigating the probability of deaths and injury in a high pressure experimental supercritical extraction apparatus in the case of explosion of the extractor vessel.

Because high pressure systems like experimental apparatus for extraction with supercritical CO_2 are now commonly used in many Research Centers and Universities, safety rules have been defined to assure that the pressure vessels, as well as the main components of a high pressure unit, are perfectly safe for continuous operation.

In a previous study, the optimal conditions and the minimum energy required for the deacidification of vegetable oils using carbon dioxide above its supercritical conditions were investigated (Penedo and Coelho, 1997). Figure 1 shows the Temperature-entropy (T,s) diagram, which summarizes both situations using typical thermodynamic data. The arrows indicate the path of the carbon dioxide in the process cycle. The extraction process involves an isentropic compression step in P_1 (1 - 2), an isobaric heating step in W_1 (2 - 3), an isenthalpic adiabatic expansion step (3 - 4), a vaporization step of the liquid phase (4 - 5) and a condensation step in W_2 that allows the use of a pump (5 - 1).

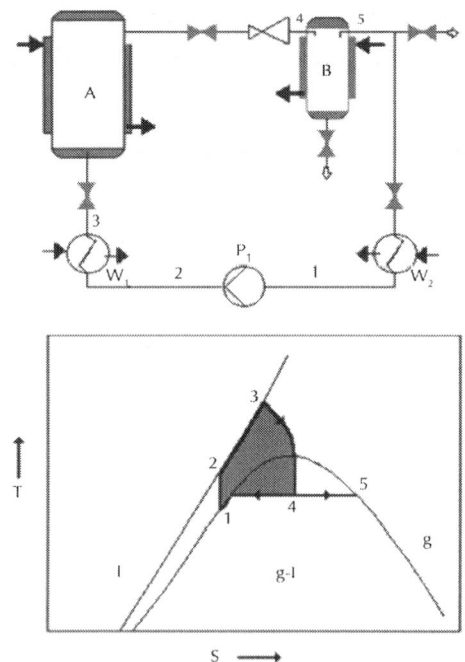

Figure 1: Flow diagram of an experimental apparatus for extraction and the temperature-entropy diagram for carbon dioxide. A and B, pressure vessel; W_1 and W_2, heat exchanger.

The energy involved in the extraction step that occurs at 353 K and 140 bar is 3.57 kJ and was read directly from the diagram, which is particularly appropriate since heat energies supplied or removed in reversible processes can be read off as areas. Assuming that 40% of this released energy is converted into overpressure and that 1.28 is the ratio of the specific heat of carbon dioxide at the critical temperature (Smith *et al.*, 2005), 1.01325 bar is the atmospheric pressure and 0.1 m³ is the initial volume of vapor present at the moment of the collapse of the vessel, then the overpressure generated is 1.1820 bar. The value of the overpressure was obtained using a numerical method together with Equation (3).

If the pressure term (P) in Equations (1) and (2) is isolated and the probit values (Pr) in Table 1 are employed, it is possible to construct a graph that associates the overpressure with the probability of death from lung injury and injury from eardrum rupture. The probit method results for explosion vulnerability are shown in Fig. 2.

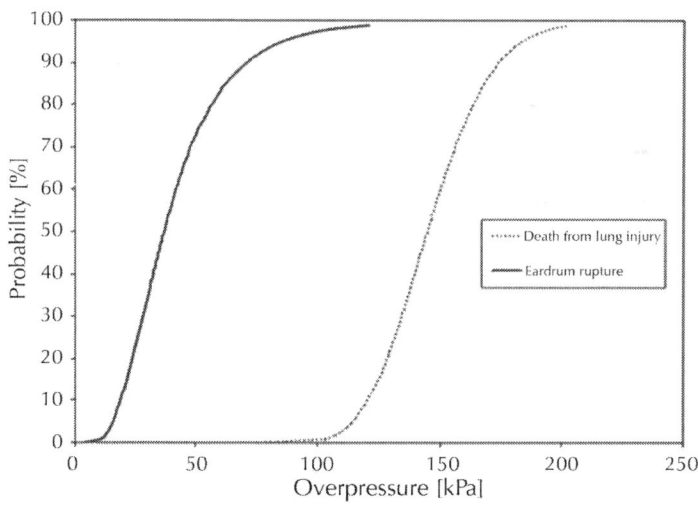

Figure 2: Relationship between probability of affected people and overpressure.

According to Figure 2, when 0.1 m³ of vapor is present in the extractor vessel in the moment of collapse of the equipment, an overpressure of 118.20 KPa (1.1820 bar) is generated. The probability that this pressure wave could cause death from lung injury is less than 1%. However, the probability that one person could have an eardrum rupture is approximately 95%, which indicates that more attention must be given to the use of protective headsets by operators of the extractor vessel under the conditions addressed in this study.

The methodology discussed in this paper can also be extended to a larger scale (pilot or industrial) by making the appropriate substitutions of the variables in the models presented. It is recommended that more accurate calculations be made for the determination of the volume of the vapor phase, as well as for the process of expansion of the gases and their actual behavior. Other effects such as thermal radiation and the projection of fragments should also be further studied. Although this case study was related to a supercritical extraction device, this methodology can be applied to any process that uses pressure vessels.

CONCLUSIONS

If pressurized gases impose extreme safety codes, due to their high compressibility, in liquids it is possible to reach high pressure values without great difficulties. In each situation it is necessary to adjust the design of the equipment and the alloy composition to the specific application. The behavior of materials under high pressure and the compressibility of liquids and gases are known in more and more detail, which allows one to design and build equipment for higher pressures and temperatures.

A safety analysis is very important in a supercritical extraction plant on an industrial, pilot or laboratory scale since high pressure is involved in the process and implies the use of pressure vessels. The explosion of an extractor vessel (pressure vessel) in a supercritical facility can cause a large release of energy, converted into overpressure, thermal radiation and kinetic energy of fragments, harming people and damaging buildings and the environment. The present study investigated the probability of deaths and injuries in a laboratory supercritical extraction unit in the case of explosion of the extractor vessel. According to the results presented, more attention should be given to the use of protective headsets because the probability of eardrum injury is superior to that of death from lung injury.

REFERENCES

1.   Abbasi, T. and Abbasi, A. S., The Boiling Liquid Expanding Vapor Explosion (BLEVE): Mechanism, consequence assessment, management. Journal of Hazardous Materials, 141, 489 (2007).

2.   Bajpai, S., Gupta, J. P., Site security for chemical process industries. Journal of Loss Prevention in the Process Industries, 18, 301 (2005).

3.   Carlès, P., A brief review of the thermophysical properties of supercritical fluids. Journal of Supercritical Fluids, 53, 11 (2010).

4.   Cavalcante, A. M., Torres, L. G. and Coelho, G. L. V., Adsorption of ethyl acetate onto modified clays and its regeneration with supercritical CO_2. Brazilian Journal of Chemical Engineering, 22, 75 (2005).

5.   Crowl, D. A. and Louvar, J. F., Chemical Process Safety: Fundamentals with Applications. Prentice Hall, New York, 426 (2002).

6.   Eggers, R., Large-Scale Industrial Plant for Extraction with Supercritical Gases. Extraction with Supercritical Gases. Verlag-Chemie, Weinheim (1980).

7.   Hendershot, D. C., An overview of inherently safer design. Process Safety Progress, 25, 13 (2006).

8.   Lucas, S., Alonso, E., Sanz, J. A. and Cocero, M. J., Safety Study

in a supercritical extraction plant. Chemical Engineering and Technology, 26, 449 (2003).

9. McHugh, M. A. and Krukonis, V. J., Supercritical Fluid Extraction: Principles and Practices. Butterworth, New York (1986).

10. Medina, H., Arnaldo, J. and Casal, J., Process Design optimization and risk analysis. Journal of Loss Prevention in the Process Industries, 22, 566 (2009).

11. Penedo, P. L. M. and Coelho, G. L. V., Optimization of deacidification of vegetable oils using supercritical CO_2. Proceedings of 4th International Symposium on Supercritical Fluids, Japan, 503 (1997).

12. Perry, H. R., Green, D. W. and Maloney, J. O., Perry's Chemical Engineer Handbook. McGraw-Hill, New York, (2008).

13. Planas-Cuchi, E., Salla, J. M. and Casal, J., Calculating overpressure from BLEVE explosions. Journal of Loss Prevention in the Process Industries, 17, 431 (2004).

14. Prugh, R. W., Quantify BLEVE hazards. Chemical Engineering Progress, 14, 66 (1991).

15. Radomski, M. R. and Ros, Z., Design of high pressure vessel used in CIP and HIP technologies. High Pressure and Biotechnology, Colloque INSERM/John Libbey Eurotext Ltd., 224, 541 (1992).

16. Ronza, A., Touza, L. L., Carol, S. and Casal, J., Economic valuation of damages originated by major accidents in port areas. Journal of Loss Prevention in the Process Industries, 22, 639 (2009).

17. Salzano, E. and Cozzani, V., A fuzzy set analysis to estimate loss intensity following blast wave interaction with process equipment. Journal of Loss Prevention in the Process Industries, 19, 343 (2006).

18. Schneider, G. M., Physicochemical aspects of fluid extraction. Fluid Phase Equilibria, 10, 141 (1983).

19. Sierra, E. T., Models of vulnerability of people and major

accidents: PROBIT method. Ministry of Labour and Social Affairs of Spain, (1991).

20. Sklet, S., Hydrocarbon releases on oil and gas production platforms: Release scenarios and safety barriers. Journal of Loss Prevention in the Process Industries, 19, 481 (2006).

21. Smith, R., van Ness, H. C. and Abbott, M. M., Introduction to Chemical Engineering Thermo-dynamics. McGraw-Hill, New York (2005).

22. Velasco, R. J., Villada, H. S. and Carrera, J. E., Applications of supercritical fluids in the agroindustry. Information Technology, 18, 53 (2007).

Tool Support for the Management of Design Processes in Chemical Engineering

Manfred Nagl[a], Bernhard Westfechtel[a], and
Ralph Schneider[b]

[a]RWTII Aachen, Lehrstuhl fu¨r Informatik III, D-52056 Aachen,
Germany
[b]RWTH Aachen, Lehrstuhl fu¨r Prozesstechnik, D-52056 Aachen,
Germany

ABSTRACT

Design processes in chemical engineering are hard to support. In particular, this applies to conceptual design and basic engineering, in which the fundamental decisions concerning the plant design are performed. The design process is highly creative, many design alternatives are explored, and both unexpected and planned feedback occurs frequently. As a consequence, it is inherently difficult to manage design processes, i.e. to coordinate the effort of experts working on tasks such as creation of flow diagrams, steady-state and dynamic simulations, etc. On the other hand, proper management is crucial because of the large economic impact of the performed design decisions. We present a management system which takes the difficulties mentioned above into account by supporting the coordination of dynamic design processes. The management system equally covers products, activities, and resources, and their mutual relationships. With respect to coverage and integration, and with respect to the dynamics of design processes, the functionality of the management system goes considerably beyond commercial project, document, and workflow management systems.

INTRODUCTION AND MOTIVATION

Design processes in engineering disciplines, like chemical engineering, often deliver good results, but sometimes perform less effective than they could. Reasons, among others, are that neither the process is explicitly and clearly structured nor its complex result. Especially, the experience of designers is not explicitly gathered and, therefore, cannot be used, the many mutual relationships between parts of the design product are neither explicitly stored nor maintained, designers may interpret the design in different ways, and the management of a project is not given a clear view about the state of the project at a certain time.

Correspondingly, there is a lack of semantic support by current tools for collaborative design processes, carried out by different persons, with different roles, on different sites, eventually in different companies, which altogether build up or maintain the complex product of a design process. Moreover, there are lot ofgaps with respect to tool support in a collaborative design process. As a consequence, the vision of an integrated design environment providing high-level tools has been realized only to a limited extent. The Collaborative Research Center476IMPROVE (Nagl and Westfechtel, 1998 and Marquardt and Nagl, 1999), is an integrated project staffed by chemical engineers, plastic engineers, ergonomics researchers, and different groups from computer science (software engineering, information systems, communication systems) with the aim of solving the problems described above. The project concentrates on the early phases of process engineering, namely conceptual process design and basic engineering, the tasks of which are to structure, to simulate, and to evaluate a plant design under different perspectives. This part of the overall design process is especially challenging from a research perspective (many creative decisions, permanent changes, study of variants etc.).

With respect to the development of an integrated design environment, IMPROVE follows a mixed top–down/bottom–up approach (Fig. 1). Bottom–up means that existing tools and platforms are re-used as far as possible (grey regions). For example, we make use of existing tools for performing steady-state or dynamic simulations. To further improve the functionality offered to the designers, new tools and services are added. The tools are designed in such a way that they fit into the overall architecture (top–down approach).

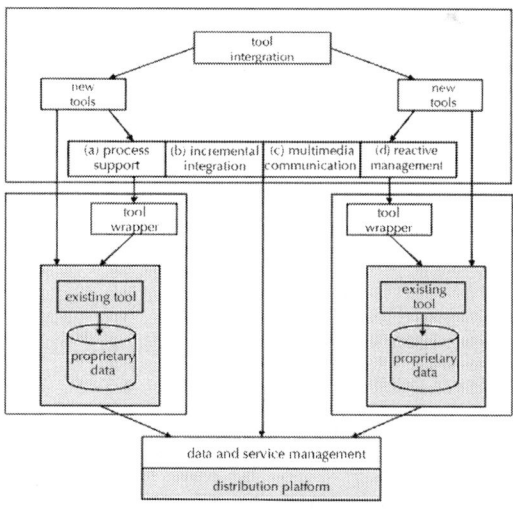

Figure 1: Mixed tool integration approach (coarse architectural sketch).

Within the IMPROVE project, there are four subprojects which address improved tool support in different functional areas. Within these subprojects, tool services are developed which may be used for the implementation of design tools with innovative functionality (Fig. 1(a)–(d)):

- Fine-grained process support tools (Pohl et al., 1999) aim at supporting designers in their interaction with design tools by process fragments which (partly) automate sequences of command invocations.

- Incremental integration tools (Becker, Haase, Westfechtel, & Wilhelms, 2002) assist designers in keeping inter-dependent design data consistent with each other (e.g., flow diagrams and simulation models).

- Using multi-media communication tools (Schüppen, Trossen, & Wallbaum, 2002), designers may discuss and resolve design problems in joint working sessions (conferences).

- Reactive management tools Westfechtel, 1999 assist in managing design processes at a more coarse-grained level than the process support tools mentioned above.

These new functionalities work synergetically together. To take one example, let us regard the change of a flow diagram, for which process support (a) can be used. The dependent tasks are determined by reactive management (d). For changing the dependent documents (e.g., a simulator input description), integration tools (b) are applied. During the change subprocess spontaneous multimedia communication is used as well as a multimedia conference (c) at the end of the subprocess. In this paper we concentrate on tool support for the management of design processes in chemical engineering (i.e. (d) in Fig. 1). Management, thereby, is restricted to the coordination of a design project. So, no strategic or psychological aspects are discussed. However, all aspects of coordination (products, activities, and resources) are seen as being tightly integrated. Since the design process changes permanently, coordination has to react accordingly (dynamics). Management is supported by a collection of tools which constitute a reactive management system. This system is called Adaptable and

Human-Centered Environment for the MAnagement of Development Processes (AHEAD Jäger, Schleicher, and Westfechtel, 1999a and Westfechtel, 1999).

Management in the sense of this paper is closely related to major design decisions of a chief designer. So, e.g., choosing one of different variants for a part of the plant induces a specific form of the design result (how many and which documents), design process (which design tasks) and corresponding required resources. Conversely, a limit in the design process' costs influences how intensively design variants can be studied. So, both levels, major design decisions on the technical level and coordination of the design process on the managerial level, are mutually dependent.

The rest of this paper is structured as follows: In Section 2, we elaborate on the features of design processes in chemical engineering. We also introduce a case study, namely the design of a plant for polyamide6. In Section 3, we discuss the management of design processes and the current state of the art of tool support. We demonstrate that commercial tools for workflow, project, or document management suffer from several shortcomings, e.g., limited support for dynamically evolving design processes. In Section 4, we present our management system (AHEAD) and explain in what respects it goes beyond the functionality of current commercial systems. In Section 5, we demonstrate its use by applying it to the case study introduced in Section 2. While the management system incorporates novel functionality, it is a research prototype which cannot be immediately applied to industrial practice. Therefore, we indicate different ways of technology transfer in Section 6. Finally, we summarize our contributions and describe current and future work in Section 7.

DESIGN PROCESSES IN CHEMICAL ENGINEERING

From an economic point of view the early phases of design processes, namely conceptual process design and basic engineering, are worth considering. The decisions made in these phases have a great impact on the later ones. McGuire and Jones (1989) report that up to 80% of the capital costs of a plant are determined in conceptual process design. The importance of design processes in (chemical) engineering and the need for improving and modeling them has been stated in the literature by several authors (e.g.,Mostow, 1985, Bañares-Alcántara, 1995, Ponton, 1995 and Westerberg, Subrahmanian, Reich, Konda, and n-dim group, 1997).

Characteristics of Design Processes

The life cycle of a chemical design process ranges from the definition of a design objective to the construction of the plant as shown in Fig. 2 (developed within the Global CAPE-OPEN projectBraunschweig & Gani, 2002).

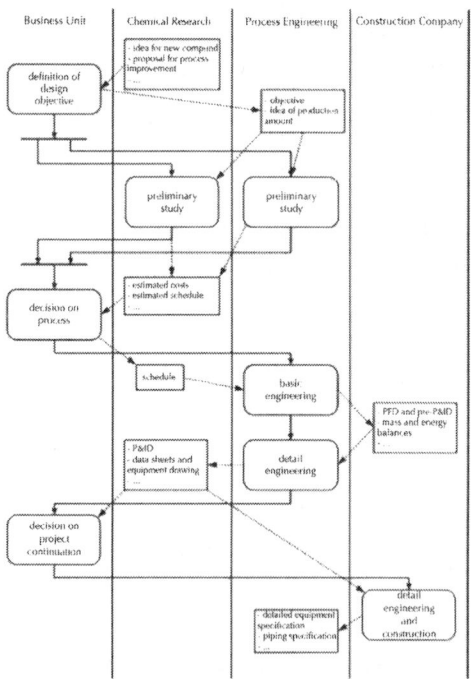

Figure 2: Life cycle of a chemical design process.

Besides the performed activities (e.g., basic engineering), which are assigned to roles or departments (resources), the needed and produced information (products) characterize the design process and are represented in the figure. The temporal order of the activities is given by the control flow (solid lines), the flow of information by the dashed lines.

Our management system aims at supporting the engineering design process (namely the activities from the start of the project to the ones including the detailed process design) with special consideration for the characteristics of a (chemical) engineering design process:

- Designers: The designers have different backgrounds and work together in a relatively small, but interdisciplinary team (up to 10 people), lead by one technical project manager.

- Location: The design team can be located in different departments and enterprises, distributed across different sites all over the world.

- Duration: Chemical engineering design projects differ in their duration, depending on the goal of the design project (e.g., design of a complete new plant, or just a retrofit). They last from several months up to many years.

- Creativity: Design processes include highly creative work processes. They are ill-defined and very complex, therefore very difficult to plan in advance.

- Iterations: There are a lot of iterations in a design process. Many activities have to be performed several times at different levels of detail (granularity) depending on the available information.

- Feedback: The activities performed during a design process are highly interconnected. Therefore changes made in a later phase of the design process can lead to unwanted feedback to activities already performed earlier during process design.

- Documentation: The exchange of information between the designers is mainly done during project meetings and by the exchange of email and paper. There is almost no reuse of successful problem solutions of previous design projects due to the difficulties in exchanging and using consistent information as well as missing or bad documentation.

- Alternatives: During a design process many different alternatives are created. The detailed investigation of each alternative may lead to totally different design processes.

- Software tools: A multiplicity of software tools is used by the designers. These tools often have incompatible data formats which make the exchange of information very difficult or even impossible. Large amounts of data and documents are produced by the software tools which have to be managed.

- Design product: The product of a design process in chemical engineering is the plant design. Chemical plants are unique and no bulk products.

In particular, these characteristics imply that design processes cannot be planned in detail in advance because they are dynamically changing due to the highly creative character of the processes and the interdependencies between the different activities and products. Effective support for managing design processes can only be enabled by the combined consideration of all these aspects.

Polyamide6 Case Study

The case study given here serves different purposes. First of all, we want to understand the workflow of industrial design processes in order to identify weak points and define requirements for the development of new tool functionalities respectively new tools. This case study can therefore be seen as a guideline for our tool design process. It is a common basis for our work in the IMPROVE project. All tools developed in IMPROVE are tested in the context of the case study. Furthermore, these tools are integrated in a common prototype demonstrating the interactions between the different tools and their support functionalities. Following this working procedure it is possible to evaluate whether the tools really fulfill the defined requirements and contribute significantly to an improvement of design processes in chemical engineering. In order to demonstrate and evaluate the functionalities of the management system, a realistic case studyis needed. Since such a case study is not available in the literature, we developed one by ourselves.

First of all we tried to record and structure design processes in chemical engineering at a very coarse level (independently from a concrete example) based on literature (Rudd, Powers, and Siirola, 1973, Douglas, 1988, Biegler, Grossmann, and Westerberg, 1997 and Blass, 1997), the PIEBASE activity model PIEBASE Working Group 2, (1998), and own experiences. Furthermore, we developed a process for the production of polyamide6 (nylon6) fulfilling a specified design task. During this development we observed ourselves and recorded our activities in the form of activity diagrams. In parallel to this we conducted five interviews with project managers of an industrial partner. In these interviews the interviewee had to remember in the sense of a case study approach (Yin, 1984) a past design process. These three elements (literature, self observation, and interviews) form the basis on which our case study is built.

The case study describes the design process of a chemical plant for the production of polyamide6 including the polymer compounding and post-processing. It focuses on the workflow, the people involved and the tools used together with their interactions. The task given in this case study is to design a plant for producing 40 000 tons polyamide6 per year with a given product quality. Polyamide6 is produced by the polymerization of -caprolactam. There are two possible reaction mechanisms, the hydrolytic and the anionic polymerization (Kohan, 1995). Since the anionic polymerization is mainly used for special polymers, this case study focuses on the hydrolytic polymerization, which is also more often applied industrially. This polymerization consists of three single reaction steps: ring opening of -caprolactam, poly-condensation, and poly-addition. The case study covers the design of the reaction and separation system as well as the extrusion. A block flow diagram representing the continuous polymerization of polyamide6 is shown in Fig. 3.

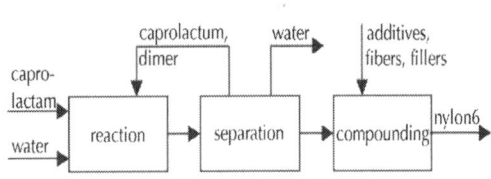

Figure 3: Block flow diagram of polyamides6 process.

There are different kinds of reactors that can be used for the polymerization: sequences of two or more tank reactors or plug flow reactors (Gerrens, 1981) and a special reactor developed for the polyamide6 production, the VK-tube (Deutsches Patentamt, 1969). The polymer melt at the outlet of the reactor contains monomer, oligomers, and water, which must be removed in order to meet the required product quality. Therefore, a separation is needed. Two separation mechanisms can be used here: evaporation in a wiped-film-evaporator or extraction with water to remove -caprolactam with a successive drying step (Kohan, 1995). Polymer post-processing is done within an extruder: additives, fillers, and fibers are added in order to meet the specified product qualities. Within the extruder, an additional polymerization of the melt and the degassing of volatile components are possible.

For all process units, mathematical models are developed and used for simulation within different simulators. Because the design of each distinct process unit requires very specific knowledge, each block is studied in detail by different experts in our case study.

The different alternatives for reaction, separation, and extrusion lead to several alternative processes for the polyamide6 production. In Fig. 4, a flow diagram of one process alternative is given: the reaction takes place within two reactors; separation is done by leaching and drying of polymer pellets. The cleaned pellets are re-melted in the extruder so that additives can be added. More detailed descriptions of the case study can be found in Bayer, Eggersmann, Gani, and Schneider (2002) and Eggersmann, Hackenberg, Marquardt, and Cameron (2002).

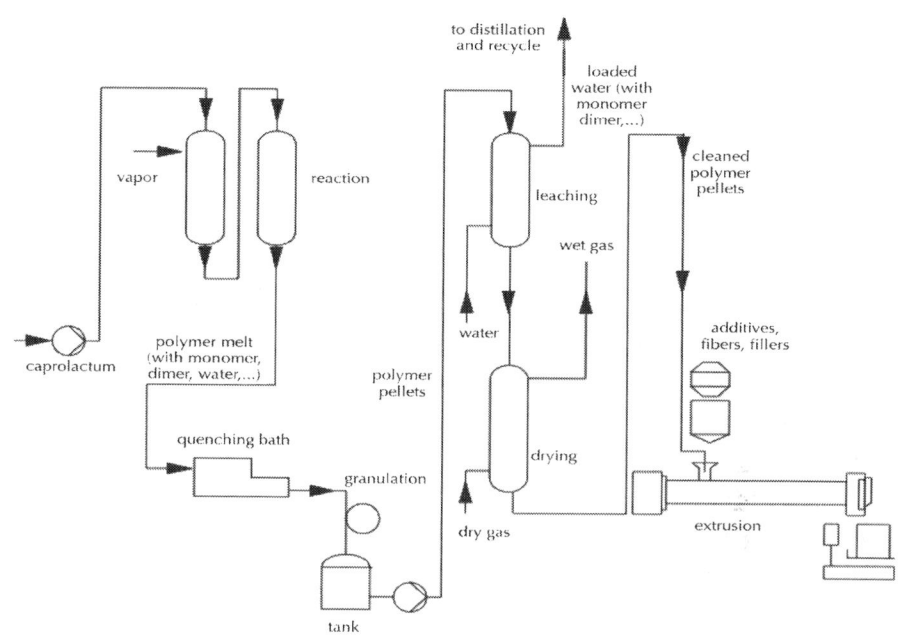

Figure 4: Flow diagram for polyamide6 production.

In Fig. 5, a simplified overview on the polyamide6 case study is given from a workflow perspective. Regarding the above mentioned applications of our case study it was important to choose a suitable notation, in the sense that engineer as well as computer scientists are able to understand the modeled content and that all necessary information is included in such a workflow model. The notation used is theC3 formalism (Foltz, Killich, Wolf, Schmidt, & Luczak, 2001), a modeling language for the notation of work processes, based on the Unified Modeling Language (uml) (Booch, Rumbaugh, & Jacobson, 1999). The abbreviation C3 stands for the three aspects of workflow modeling which are represented in this formalism: cooperation, coordination, and communication. The elements of C3 are roles (e.g., simulation expert), activities (e.g., design reaction alternatives), input/output information (not shown in this figure), control flows (solid lines including forks and joins represented by bars), information flows (also not shown), and synchronous communication (represented by a filled square).

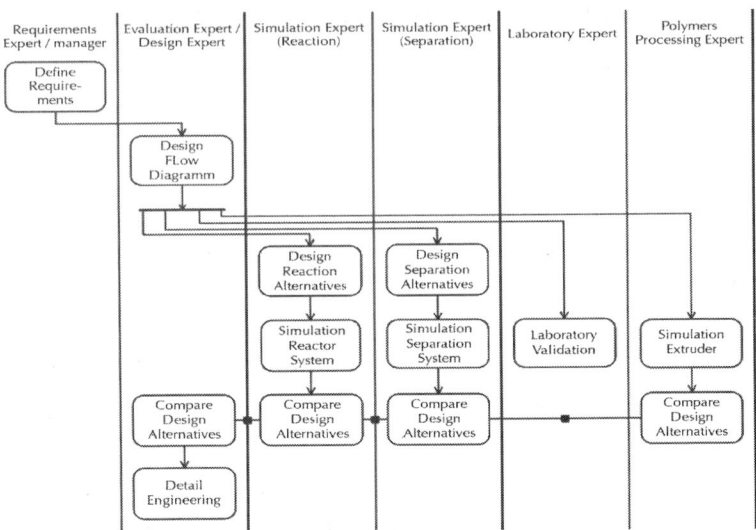

Figure 5: Simplified overview on case study.

After the start of the project, different alternatives are evaluated on the basis of block flow diagrams. Reaction, separation, and extrusion are investigated in parallel within different simulation tools. The simulation results are compared with experimental results leading to an improvement and refinement of the used mathematical models. After the completion of these activities a plant concept is determined. The case study in its actual form represents one part of an industrial design process, namely the early phases of basic engineering.

On the basis of this case study (with about one hundred activities) the management system has been developed and tested. For reasons of clarity only a small part of the case study is presented and used for demonstration in the following sections.

MANAGEMENT OF THE DESIGN PROCESS

Basic Notions

In the previous section, we have introduced our domain of discourse—design processes in chemical engineering—and a case study (design of a plant for producing polyamide6) which serves as a reference scenario within the IMPROVE project. In this paper, we are specifically concerned with the management of design processes. In Section 2.1, we have described several features of design processes which challenge the capabilities of management. In particular, design processes are highly creative, many iterations are necessary, the tasks to be solved are not known beforehand, design processes last for a long time span, experts from different disciplines have to cooperate smoothly, these experts use heterogeneous tools, and they have to document the results of their work in a traceable way.

Before we discuss solutions to these problems, we have to introduce a set of basic notions which we will use throughout the rest of this paper. In general terms, management can be defined as 'all the activities and tasks undertaken by one or more persons for the purpose of planning and controlling the activities of others in order to achieve an objective or complete an activity that could not be achieved by the others acting alone' (Thayer, 1988). This definition stresses coordination as the essential function of management.

More specifically, we focus on the management of design processes by coordinating the technical work of designers. We do not target senior managers who work at a strategic level and are not concerned with the details of enterprise operation. Rather, we intend to support project managers who collaborate closely with the designers performing the technical work. Such managers, who are deeply involved in the operational business, need to have not only managerial but also technical skills ('chief designers').

The distinction between persons and roles is essential: when referring to a 'manager' or a 'designer', we are denoting a role, i.e. a collection of authorities and responsibilities. However, there need not be a 1:1 mapping between roles and persons playing roles. In particular, each person may play multiple roles. For example, in chemical engineering it is quite common that the same person acts both as a manager coordinating the project and as a (chief) designer who is concerned with technical engineering tasks.

In order to support managers in their coordination tasks, design processes have to be dealt with at an appropriate level of detail. We may roughly distinguish between three levels of granularity:

- At a coarse-grained level, design processes are divided into phases (or working areas) according to some life cycle model (Fig. 2).
- At a medium-grained level, design processes are decomposed further down to the level of documents or tasks, i.e. units of work distribution.
- At a fine-grained level, the specific details of design subprocesses are considered. For example, a simulation expert may build up a simulation model from mathematical equations.

Given our understanding of management as explained above, the coarse-grained level does not suffice; rather, decomposition has to be extended to the medium-grained level. On the other hand, management is usually not interested in the technical details of how documents are structured or how the corresponding personal subprocess is performed. Thus, the managerial level, which defines how management views design processes, comprises both coarse- and medium-grained representations. In order to support managers in their coordination tasks, they must be supplied with appropriate views (abstractions) of design processes. Such views must be comprehensive inasmuch as they include products, activities, and resources (and their mutual relationships):

- The term product denotes the results of design subprocesses (e.g., flow diagrams, simulation models, simulation results, cost estimates, etc.) [1]. These may be organized into documents, i.e. logical units which are also used for work distribution or version control.

- The term activity denotes an action performing a certain function in a design process. At the managerial level, we are concerned with tasks, i.e. descriptions of activities assigned to designers by managers.

- Finally, the term resource denotes any asset needed by an activity to be performed. This comprises both human and computer resources (i.e. the designers and managers participating in the design process as well as the computers and the tools they are using).

Thus, an overall management configuration consists of multiple parts representing products, activities, and resources. An example is given in Fig. 6. Here, we refer to the polyamide6 design process introduced earlier. On the left, the figure displays the roles in the design team as well as the designers filling these roles [2]. The top region on the right shows design activities connected by control and data flows. Finally, the (versioned) products of these activities are located in the bottom–right region.

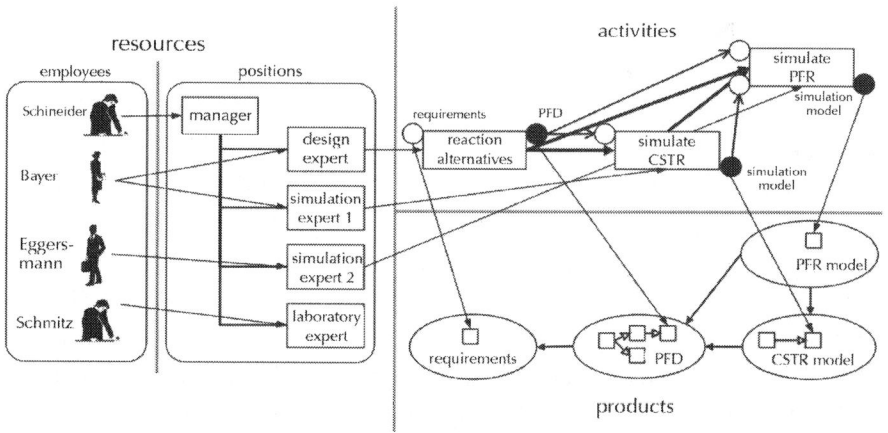

Figure 6: Management configuration.

Below, we give a more detailed description of Fig. 6:

- Products. The results of design processes such as process flow diagrams (PFDs), steady-state and dynamic simulations, etc. are represented by documents (ellipses). Documents are interdependent, e.g., a simulation model depends on the PFD to which it refers (arrows between ellipses). The evolution of documents is captured by version control (each box within an ellipsis represents a version of some document).

- Activities. The overall design process is decomposed into tasks (rectangular boxes) which have inputs and outputs (white and black circles, respectively). The order of tasks is defined by control flows (thick arrows); e.g., reaction alternatives must have been inserted into the flow diagram before they can be simulated. Finally, data flows (arrows connecting circles) are used to transmit document versions from one task to the next.

- Resources. Employees (icons on the left) such as Schneider, Bayer, etc. are organized into project teams which are represented by organization charts. Each box represents a position, lines reflect the organizational hierarchy. Employees are assigned to positions (or roles). Within a project, an employee may play multiple roles. e.g., Mrs Bayer acts both as a designer and as a simulation expert in the polyamide6 team.

- Integration. There are several relationships between products, activities, and resources. In particular, tasks are assigned to positions (and thus indirectly to employees). Furthermore, document versions are created as outputs and used as inputs of tasks.

It is crucial to understand the scope of the term 'management' as it is used in this paper. As already stated briefly above, management requires a certain amount of abstraction. This means that the details of thetechnical level are not represented at the managerial level. This is illustrated in Fig. 7, whose upper part shows a small cutout of the management configuration of Fig. 6. On the managerial level, the design process is decomposed into activities such as creation of reaction alternatives and simulation of these alternatives. Activities generate results which are stored in document versions. At the managerial level, these versions are basically considered black boxes, i.e. they are represented by a set of descriptive attributes (author, creation date, etc.) and by references to the actual contents, e.g., PFDs and simulation models. How a PFD or a simulation model is structured internally (and how their contents are related to each other), goes beyond the scope of the managerial level. Likewise, the managerial level is not concerned with the detail personal process which is executed by some human to create a PFD, a simulation model, etc.

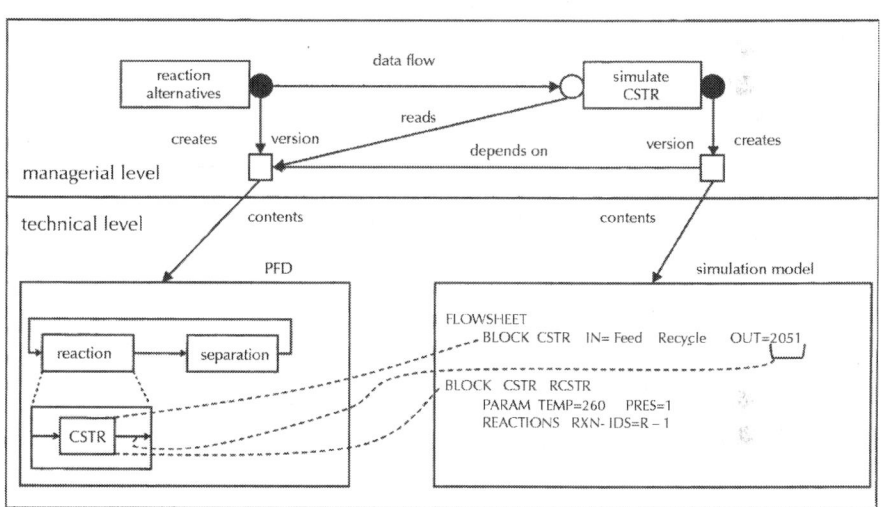

Figure 7: Managerial and technical level.

This does not imply that technical details are ignored. Rather, it must be ensured that the managerial level actually constitutes a correct abstraction of the fine-grained information at the technical level—and also controls technical activities. In fact, the management system described in this paper is part of an integrated environment for supporting design processes in chemical engineering (see also Fig. 1). As such, it is integrated with tools providing fine-grained product and process support (Bayer et al., 2002; Becker et al., 2002). The interplay of the tools of the overall environment is sketched only briefly in this paper; see (Nagl, Schneider, & Westfechtel, in press).

Particular attention has to be paid to the dynamics of design processes (called evolution in (Smithers & Troxell, 1990)). As we have demonstrated in the previous section, the design process is not known in advance. Rather, it continuously evolves during execution. Often, the term 'dynamics' is interpreted in a rather restricted way, referring only to the execution of a known process with a static definition. In addition, we have to take care of evolving definitions, i.e. the activities to be executed as well as their relationships are defined only at runtime.

As a consequence, all parts of a management configuration evolve continuously:

- Products. The product structure is determined only during the design process. It depends on the flow diagram which is continuously extended and modified. Other documents such as simulation models and simulation results depend on the flow diagram. Moreover, different variants of the chemical process are elaborated, and selections among them are performed according to feedback gained by simulation and experiments.

- Activities. The activities to be performed depend on the product structure; feedback may require the re-execution of terminated activities, concurrent/simultaneous engineering calls for sophisticated coordination of related activities, etc.

- Resources. Resource evolution occurs likewise: new tools arrive, old tool versions are replaced with new ones, the project team may shrink due to budget constraints, or it may be extended to meet a crucial deadline, etc.

However, a management configuration should not evolve in arbitrary ways. There are domain-specific constraints which have to be met. In particular, activities can be classified into types such as requirements definition, design, simulation, etc. (likewise for products and resources). Furthermore, the way how activities are connected is constrained as well. For example, a flow diagram can be designed only after the requirements have been defined. Such domain-specific constraints should be taken into account such that they restrict the freedom of evolution.

Management Systems: State of the Art

Above, we have discussed design processes and their management on a conceptual level. In the following, we will be concerned with tool support for managing design processes. From the previous discussion, we derive a set of crucial requirements for management tools for design processes [3]:

- Medium-grained representation. The management of design processes has to be supported at an appropriate level of detail. As we have argued before, this requires a medium-grained representation of the design process.

- Coverage and integration at the managerial level. Management tools have to deal equally with products, activities, and resources. In addition, the relationships among them have to be taken into account.

- Integration between managerial and technical level. Managerial activities have to be coupled with technical activities. Practically speaking, this implies e.g., that designers have to be supplied with the documents they are going to manipulate, as well as with the tools they are going to use.

- Dynamics of design processes. Design processes evolve continuously during execution. Thus, a management system must support dynamic changes so that product evolution, feedback, simultaneous and concurrent engineering etc. can be expressed.

- Adaptability. Management tools have to be adapted to a specific application domain. For example, they must be aware of the types of activities performed in design processes in chemical engineering, and they must provide domain-specific operations to their users.

In industry, a large variety of commercial systems are being used for the management of design processes. These include systems for project management, workflow management, and product management, which are discussed in turn below. All of these systems meet the requirements stated above only partially (Table 1[4]).

Table 1: Comparison of AHEAD with commercial management systems

	AHEAD	Project management systems	Workflow management systems	Product management systems
Granularity of representation	Medium grained	Coarse-grained	Medium- and fine-grained	Medium-grained
Coverage at the managerial level	Products, activities, resources	Activities, resources	Activities, resources	Products, (activities)
Integration with technical level	Tool integration, document storage	Not supported	Tool integration	Document storage
Support for dynamic design processes	Full support of process evolution	Evolving project plans	Limited (fixed workflows)	Version control for documents
Adaptability	uml models	Not supported	Workflow definitions	Database schema

Project management systems (Kerzner, 1998) such as e.g., Microsoft Project support management functions such as planning, organizing, monitoring, and controlling. The project plan acts as the central document which may be represented in different ways, e.g., as a PERT or GANTT chart. It defines the milestones to be accomplished and provides the foundation for scheduling of resource utilization as well as for cost estimation and control. Project management systems are widely used in practice, but they still suffer from several limitations: project plans are often too coarse-grained, products (documents) are not considered, project plans are not integrated with the actual work performed by engineers, and there is no way to define domain-specific types of project plans. Workflow management systems (Jablonski and Bußler, 1996 and Lawrence, 1997), e.g., Staffware, FlowMark, or COSA, have been applied in banks, insurance companies, administrations, etc. A workflow management system manages the flow of work between participants, according to a defined procedure consisting of a number of tasks (McCarthy & Bluestein, 1991). It coordinates user and system participants to achieve defined objectives by set deadlines. To this end, tasks and documents are passed from participant to participant in a correct order. Moreover, a workflow management system may offer an interface to invoke a tool on a document either interactively or automatically. Their most important restriction is limited support for the dynamics of design processes. Many workflow management systems assume a statically defined workflow that cannot be changed during execution. In this way, dynamic design processes can be supported only to a limited extent (i.e. the statically known fractions can be handled by the workflow management system). Recently, this problem has been addressed in a few university prototypes (see e.g., Derniame, Baba, and Wastell, 1998 and Georgakopoulos, Prinz, and Wolf, 1999).

In the context of this paper, we use the term product management system to refer to all kinds of systems for storing, manipulating, and retrieving the results of design processes. Depending on the context in which they are employed, they are called engineering data management systems, product data management systems (Harris, 1996), software configuration management systems (Tichy, 1994 and Whitgift, 1991), or document management systems. Documentum and Matrix One are examples of such systems which are used in chemical engineering. Documents such as flow diagrams, steady-state and dynamic simulation models, cost estimations, etc. are stored in a database which records the evolution of documents (i.e. their versions) and aggregates them into configurations. In addition, product management systems may offer simple support for the management of activities (e.g., change request processes based on finite state machines), or they may include workflow components, which suffer from the restrictions already discussed above. Their primary focus still lies on the management of products; in particular, management of human resources is hardly considered. All of the approaches cited above are domain-independent. For example, workflow management systems may be applied to arbitrary business processes, and product management systems may be used in different engineering disciplines. We are aware of only a few domain-specific approaches which have been developed for chemical engineering. For example, n-dim (Levy et al., 1993; Westerberg et al., 1997) is a distributed and collaborative computer-aided environment for process engineering design; KBDS (Bañares-Alcántara & Lababidi, 1995) deals with the management of design alternatives and design histories. These approaches are better tailored towards design processes in chemical engineering. However, they do not provide comprehensive support for the management of products, activities, and resources. Moreover, they lack the generality of domain-independent systems which can be used in and adapted to different domains.

A MANAGEMENT SYSTEM FOR DE-SIGN PROCESSES

Since current management systems suffer from several limitations explained in the previous section, we have designed and implemented a new management system which addresses these limitations. This system is called AHEAD (Jäger, Schleicher, and Westfechtel, 1999b and Westfechtel, 1999). AHEAD is a research prototype that goes beyond commercial systems with respect to the requirements introduced earlier (Table 1):

- Medium-grained representation. In contrast to project management systems, design processes are represented at a medium-grained level, allowing managers to effectively control the activities of designers. Management is not performed at the level of milestones; rather, it is concerned with individual tasks such as 'simulate the CSTR reactor'.

- Coverage and integration at the managerial level. AHEAD is based on an integrated management model which equally covers products, activities, and resources. In contrast, project and workflow management systems primarily focus on activities and resources, while product management systems are mainly concerned with the products of design processes.

- Integration between managerial and technical level. In contrast to project management systems, the AHEAD system also includes support tools for designers that supply them with the documents to work on, and the tools that they may use.

- Support for the dynamics of design processes. While many workflow management systems are too inflexible to allow for dynamic changes of workflows during execution, AHEAD supports evolving design processes, allowing for seamless integration of planning, execution, analysis, and monitoring.

- Adaptability. Both the structure of management configurations and the operations to manipulate them can be adapted by means of a domain-specific object-oriented model based on the uml (Booch et al., 1999).

Functionality and Concepts

Fig. 8 gives an overview of the AHEAD system. AHEAD offers environments for different kinds of users, which are called modeler, manager, and designer. In the following, we will focus on the functionality that the AHEAD system provides to its users. Its technical realization will be discussed in the next subsection.

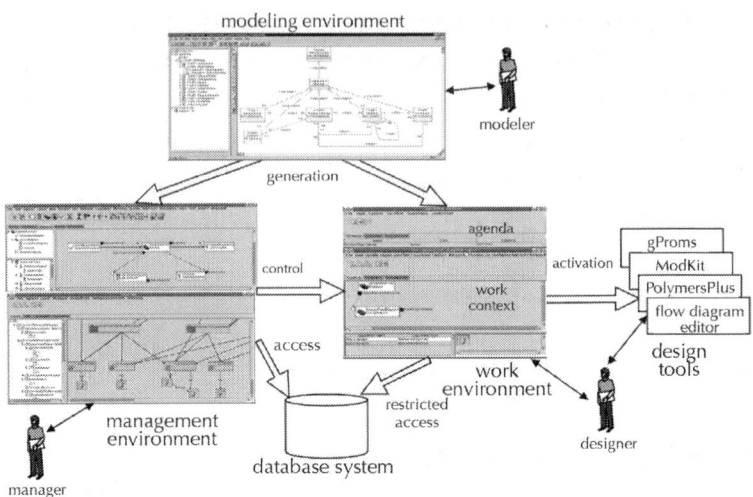

Figure 8: Architecture of the AHEAD system.

The management environment supports project managers in planning, analyzing, monitoring, and controlling design processes. It provides graphical tools for operating on management configurations. These tools address the management of activities, products, and resources, respectively (Krapp, Krüppel, Schleicher, & Westfechtel, 1998):

- For activity management, AHEAD offers dynamic task nets which allow for seamless interleaving of planning, analyzing, monitoring, and controlling. A task net consists of tasks that are connected by control flow and data flow relationships. Furthermore, feedback in the design process is represented by feedback relationships. Tasks may be decomposed into subtasks, resulting in task hierarchies. The manager constructs task nets with the help of a graphical editor. He may modify task nets at any time while a design process is being executed.

- Product management is concerned with documents such as flow diagrams, simulation models, cost estimations, etc. AHEAD offers version control for these documents with the help of version graphs. Relationships (e.g., dependencies) between documents are maintained as well. Versions of documents may be composed into configurations, thereby defining which versions are consistent with each other. The manager may view the version histories and configurations with the help of a graphical tool. In this way, he may keep track of the work results produced by the designers.

- Resource management deals with the organizational structure of the enterprise as far as it is relevant to design processes. AHEAD distinguishes between abstract resources (positions or roles) and concrete resources (employees). The manager may define a project team and then assign employees to the project positions.

Management of activities, products, and resources is fully integrated: tasks are assigned to positions, inputs and outputs of tasks refer to document versions. Moreover, AHEAD manages task-specific workspaces of documents and supports invocation of design tools (see below).

AHEAD does not only support managers. In addition, it offers a work environment which consists of two major components:

- The agenda tool displays the tasks assigned to a designer in a table containing information about state, deadline, expected duration, etc. The designer may perform operations such as starting, suspending, finishing, or aborting a task.
- The work context tool manages the documents and tools required for executing a certain task. The designer is supplied with a workspace of versioned documents. He may work on a document by starting a tool such as e.g., a flow diagram editor, a simulation tool, etc.

Please note that the scope of support provided by the work environment is limited. We do not intend to support design activities in detail at a technical level. Rather, the work environment is used to couple technical activities with management. There are other tools which support design activities at a fine-grained level. For example, a process-integrated flow diagram editor (Bayer, Marquardt, Weidenhaupt, and Jarke, 2001) may be activated from the work environment. 'Process-integrated' means that the designer is supported by process fragments which correspond to frequently occurring command sequences. These process fragments encode the design knowledge which is available at the technical level. This goes beyond the scope of the AHEAD system, but it is covered by the overall environment for supporting design processes to which AHEAD belongs as a central component.

Both the management environment and the work environment access a common management database. However, they access it in different ways, i.e., they invoke different kinds of functions. The work environment is restricted to those functions which may be invoked by a designer. The management environment provides more comprehensive access to the database. For example, the manager may modify the structure of a task net, which is not allowed for a designer.

Before the AHEAD system may be used to carry out a certain design processes, it must be adapted to the respective application domain Schleicher, 1999. AHEAD consists of a generic kernel which is domain-independent. Due to the generality of the underlying concepts, AHEAD may be applied in different domains such as software, mechanical, or chemical engineering. On the other hand, each domain has its specific constraints on design processes. The modeling environment is used to provide AHEAD with domain-specific knowledge, e.g., by defining task types for flow diagram design, steady-state and dynamic simulation, etc. From a domain-specific process model, code is generated for adapting the management and the work environment.

Realization

AHEAD is a research prototype which has been developed to demonstrate novel functionality. To implement AHEAD, we have used powerful homegrown tools for rapid prototyping (see below). This was the fastest way to obtain a demonstrator, which, however, cannot be immediately employed in industry. InSection 6, we will discuss technology transfer, which can be achieved by reusing commercial tools for project, workflow, and product management.

In the current realization of the AHEAD system, graph technology plays an important role. All data for representing management configurations are stored in the graph-based database management system GRAS (Kiesel, Schürr, & Westfechtel, 1995). Graph structures and operations are formally specified in the high-level specification language progres (Schürr, Winter, & Zundorf, 1999), which is based on programmed graph transformations. In AHEAD, the specification written in progres describes how management configurations are built up and what operations for manipulating them are offered to end users. From the specification, code is generated which operates on the management database.

For the user interface, we have developed the UPGRADE framework (Universal Platform for GRAph-Based Application DEvelopment (Böhlen, Jäger, Schleicher, & Westfechtel, 2002)). UPGRADE is implemented in Java, based on standard libraries and both public-domain and commercial components (ILOG JViews). It mainly focuses on graphical tools, but it also supports e.g., tabular and tree representations. Graphical tools provide external views on the underlying management graph, hiding all of the technical details of the internal representation.

From the work environment, external tools may be started with the help of wrappers. Wrappers constitute tool envelopes which are responsible for supplying tools with data (documents checked out from the product management database), preparing the operating system environment (e.g., by setting environment variables), and calling the tools with appropriate parameters. Wrappers are realized on top ofcorba, an infrastructure for distributed object-oriented computing. The realization of wrappers has been performed by another partner participating in the IMPROVE project (Lipperts & Thißen, 1999).

The modeling environment is realized with the help of a commercial CASE tool (Rational Rose), which is based on the uml (Booch et al., 1999). uml is a language that serves as a standard notation for object-oriented modeling. For the purpose of process modeling, we have adapted and restricted uml according to our requirements (Jäger et al., 1999b). A uml model is automatically transformed into a (part of a) progresspecification (i.e., the end user is not concerned with progres at all). This transformation tool is realized with the help of the OLE interface providing access to Rose's model database.

APPLYING THE MANAGEMENT SYSTEM TO CHEMICAL ENGINEERING

Within the IMPROVE project, the polyamide6 design process played a key role not only on a conceptual level. In addition, it was used for the development of a demonstrator integrating all software components contributed by the various project partners. The AHEAD system served as the central, coordinating component of this demonstrator (see also Section 2.2). It was integrated with the following tools: the PRIME flow diagram editor (Bayer et al., 2001), ModKit (for creating dynamic simulation models, see (Bogusch, Loehmann, & Marquardt, 2001)), KomPakt (for synchronous multi-media communication (Schüppen et al., 2002)), and Morex (for extruder design) were contributed by partners involved in the IMPROVE project; excel (from Microsoft) and PolymersPlus (from Aspen Tech) are commercial tools which are used for cost calculations and steady-state simulations, respectively. For the demonstrator, a comprehensive demo session was prepared which was successfully presented at the IMPROVE project review in May 2000 and at a national workshop with participants from chemical industry and tool vendors that was held in November 2000.

The Polyamide6 Design Process

Based on the case study introduced in Section 2.2, the polyamide6 design process is represented as a dynamic task net in Fig. 9. This task net was created from a representation in the C3 modeling language which we used in Fig. 2 and Fig. 5. Please note that Fig. 5

shows a very simplified view of the design process. To construct the task net of Fig. 9, we used a more detailed C3 model.

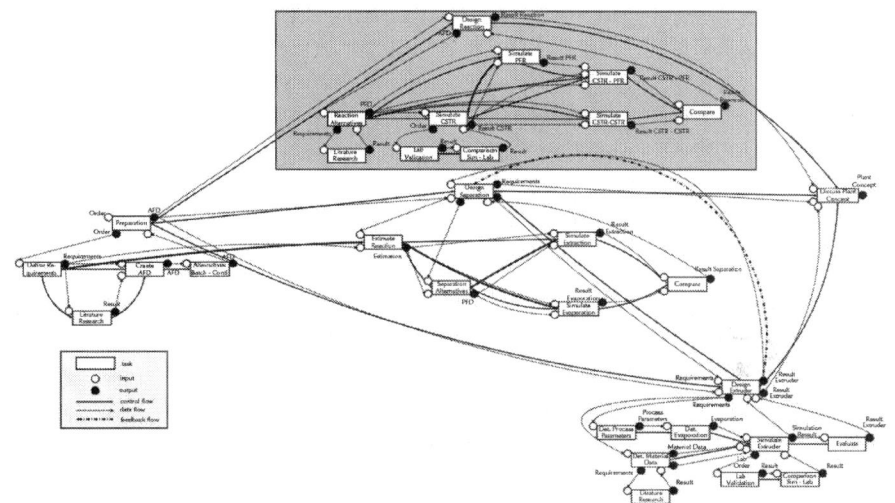

Figure 9: Task net for the polyamide6 design process.

As we will demonstrate in the demo session of Section 5.2, the task net of Fig. 9 is not available at the start of the design process. Rather, it is built up as the design proceeds. In an early phase, the task net will just contain a top-level decomposition into design tasks for reaction, separation, and compounding, respectively. However, in the IMPROVE project we were dealing with a case study which was designed beforehand (and was used to plan the development of the demonstrator). Therefore, a C3 model of the overall design process was already available in this artifical setting; it would not be available in real-world use.

The mapping from C3 to task nets was performed manually, albeit in a systematic manner. In general, human expertise is required to perform the mapping. On the other hand, there is a straightforward initial mapping which may be obtained by applying the following simple rules:

Each activity is mapped onto a task. This also applies to cooperative activities, which occur only once in a task net.[5]

- Ordering relationships between activities are mapped onto control flows. While forks and joins are represented in C3 by bars, they are not shown explicitly in task nets. Rather, they appear as bundles of outgoing and incoming control flows, respectively.
- An arrow from an activity to a document is mapped onto an output parameter of the respective task. For each arrow from a document to an activity, an input parameter is generated, and output and input are connected by a data flow.

Human expertise is required e.g., for introducing task hierarchies. The C3 model which we used as input was flat. Based on domain knowledge, reasonable subprocesses may be identified. In our case study, we have introduced subprocesses for designing the reaction, the separation, and the compounding of the chemical process. In addition, there is a preparation phase in which the overall design problem is decomposed, and a final integration phase, where the overall plant design is synthesized and discussed among the involved design experts.

Please note that reaction, separation, and compounding are designed in an integrated way:

- To design the separation, input is required with respect to incoming streams. In the sample process, this input is generated by an initial estimation which is refined later on. In this way, reaction and separation may be studied in parallel. This speeds up the overall design process.
- With respect to separation and compounding, there exist some degrees of freedom with respect to the decomposition of the overall chemical process. In particular, if separation is performed with the help of evaporation, evaporation can be performed partly in the extruder. Therefore, the respective design processes are arranged in a feedback loop for optimizing the chemical process.

In the demo session to be presented below, we focus on the design of the reaction (shaded region in Fig. 9). After an initial PFD has been created which contains multiple design variants, each of these variants is explored by means of simulations and (if required) laboratory experiments. In a final step, these alternatives are compared against each other, and the most appropriate one is selected. This simplified part of the design process suffices to demonstrate many of the essential features provided by the AHEAD system; furthermore, it is sufficiently small to be presented in this paper.

Demo Session

In this subsection, we illustrate the functionality of the AHEAD system with the help of some snapshots. We will primarily focus on the management environment; the work environment will be discussed rather briefly. The modeling environment goes beyond the scope of this paper; see Jäger et al. (1999a).

Fig. 10 presents a snapshot from the management environment taken in an early stage of the polyamide6 design process. The upper region on the left displays a tree view of the task hierarchy. The lower left region offers a view onto the resources available for task assignments (see also Fig. 11). A part of the overall task net is shown in the graph view on the right-hand side. Each task is represented by a rectangle containing its name, the position to which the task has been assigned, and an icon representing its state (e.g., the gear-wheels represent the state Active, and the hour-glass stands for the state Waiting). Black and white circles represent outputs and inputs, respectively. These are connected by data flows (thin arrows). Furthermore, the ordering of task execution is constrained by control flows (thick arrows). Hierarchical task relations (decompositions) are represented by the graphical placement of the task boxes (from top to bottom) rather than by drawing arrows (which would clutter the diagram).

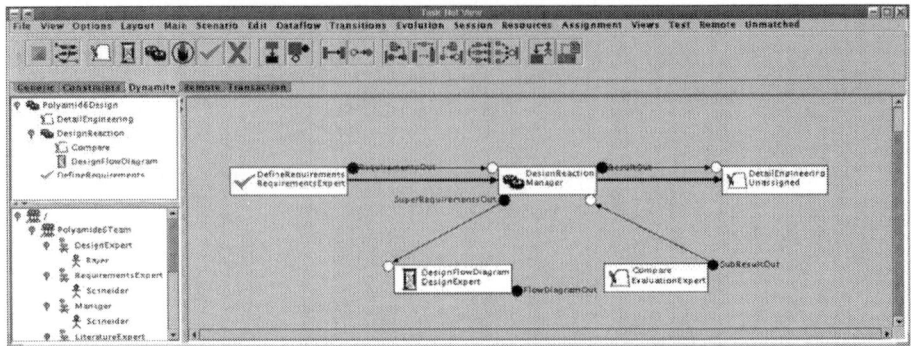

Figure 10: Initial task net (management environment).

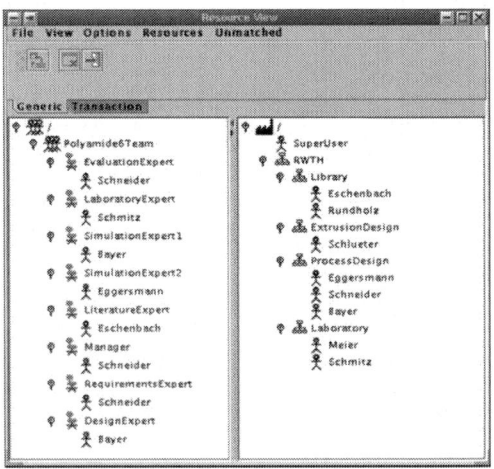

Figure 11: Resource view (management environment).

Please recall that the demo session deals only with the reaction part, i.e., we do not consider separation, extrusion, etc. On the top level of the task net of Fig. 10, the process is decomposed into three tasks: Define Requirements, Design Reaction, and Detail Engineering. Only the design of the reaction is elaborated in the sequel. In this early stage, it is only known that initially some reaction alternatives have to be designed. The result of this design task is documented in a flow diagram. Furthermore, at the end these alternatives have to be compared, and a decision has to be performed. All other tasks—e.g., for performing simulations and laboratory experiments—have to be filled in later. Thus, the initial task net is incomplete.

In addition to the initial task net, the manager has also used the resource management tool for building up his project team (Fig. 11). The region on the left displays the structure of the polyamide6 design team. Each position (represented by a chair icon) is assigned to a team member. Analogously, the region on the right shows the departments of the company. From these departments, the team members for a specific project are taken for a limited time span. The management environment offers commands for defining both the team and the department structure, and for assigning persons to positions in design teams. Please note that tasks are assigned to positions rather than to actual employees (see lower left view in Fig. 10). In this way, assignment is decomposed into two steps. The manager may assign a task to a certain position even if this position has not been filled yet. Moreover, if a different employee is assigned to a position, the task assignments need not be changed: The tasks will be redirected to the new employee automatically.

The work environment is illustrated in Fig. 12. As a first step, the user logs into the system (not shown in the figure). After that, AHEAD displays an agenda of tasks assigned to this user (more precisely: assigned to the roles played by this user). In Fig. 12, the agenda is shown in the top window. Since the user Bayer plays the role of the design expert, the agenda contains the task Design Flow Diagram. After the user has selected a task from the agenda, the work context for this task is opened (bottom window). The work context graphically represents the task, its inputs and outputs, as well as its context in the task net (here, the context includes the parent task which defines the requirements to the flow diagram to be designed). Furthermore, it displays a list of all documents needed for executing this task. For some selected document, the version history is shown on the right (so far, there is only one version of the requirements definition which acts as input for the current task).

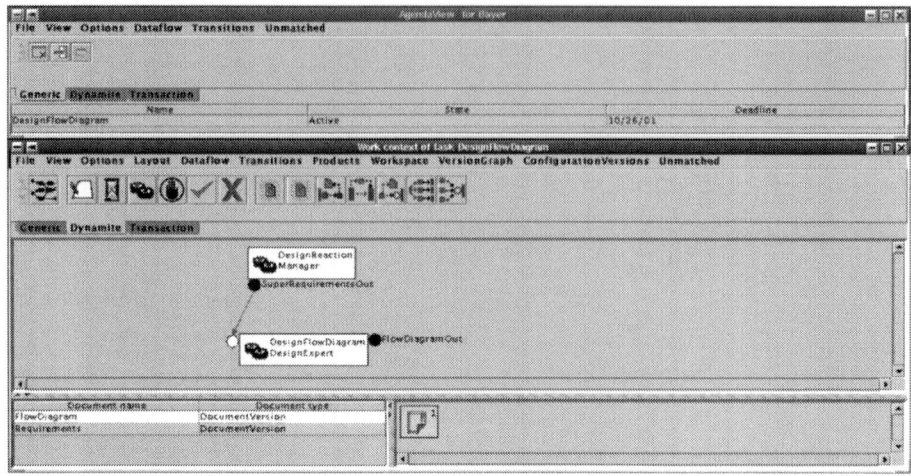

Figure 12: Work environment.

From the work context window, the user may activate design tools for operating on the documents contained in the workspace. Here, the user invokes a flow diagram editor (Bayer et al., 2001) in order to insert reaction alternatives into the flow diagram for the polyamide6 process. The flow diagram editor, which was also developed in the IMPROVE project, is based on MS Visio, a commercial drawing tool, which was integrated with the PRIME process engine prime (Pohl et al., 1999). The flow diagram editor supports hierarchical flow diagrams (abstract or process flow diagrams); furthermore, it may represent alternative refinements (variants) for blocks occurring in the flow diagram. The flow diagram editor offers fine-grained process fragments which are used for guidance and automation. These fragments incorporate domain-specific knowledge. For example, there is a process fragment which assists the designer in introducing reaction alternatives into the flow diagram (based on experience from previous design processes).

The resulting flow diagram is displayed in Fig. 13. The chemical process is decomposed into reaction, separation, and compounding. The reaction is refined into four variants. For our demo session, we assume that initially only two variants are investigated (namely a single CSTR and PFR, respectively). That is, at the current state of design the alternatives on the right-hand side have not yet been introduced into the flow diagram; they will be considered later on.

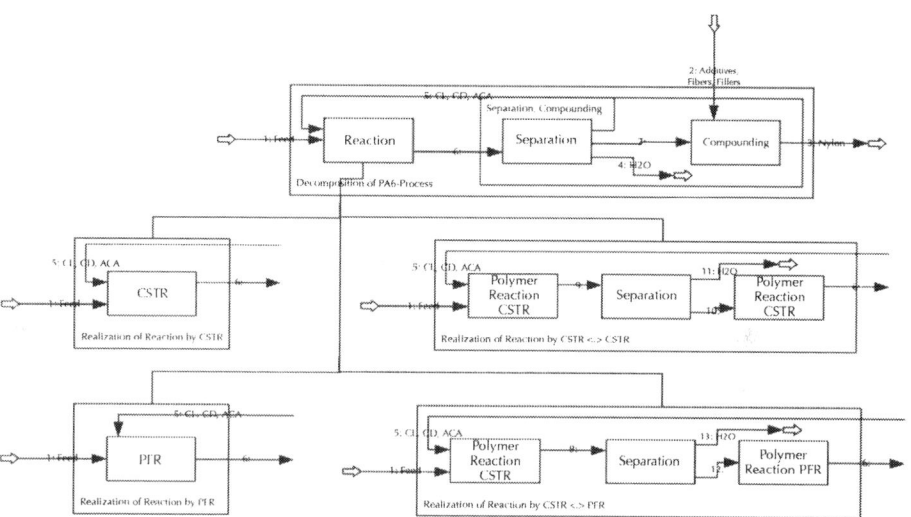

Figure 13: Reaction alternatives in the PFD.

After the generation of the four variants, the manager extends the task net with tasks for investigating the alternatives that have been introduced so far (product-dependent task net, Fig. 14). Please note the control flow relation between the new tasks: The manager has decided that the CSTR should be investigated first so that experience from this alternative may be re-used when investigating the PFR. Furthermore, we would like to emphasize that the design task is not terminated yet. As to be demonstrated below, the designer waits for feedback from simulations in order to enrich the flow diagram with simulation data. Depending on these data, it may be necessary to investigate further alternatives.

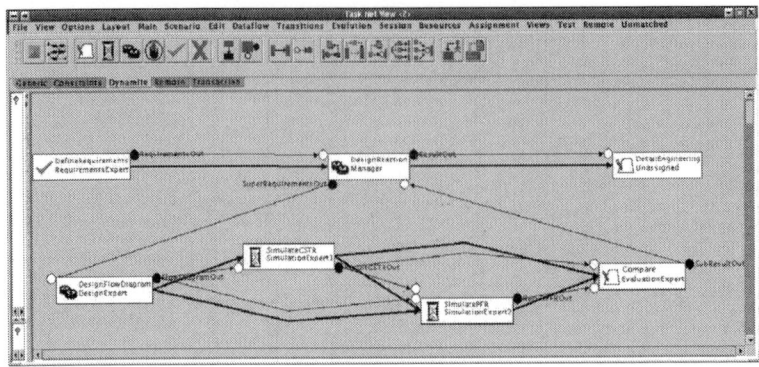

Figure 14: Extended tesk net (management environment).

Subsequently, the simulation expert creates a simulation model (using PolymersPlus) for the CSTR reactor and runs the corresponding simulations. The simulation results are validated with the help of laboratory experiments. After these investigations have been completed, the flow diagram can be enriched with simulation data such as flow rates, pressures, temperatures, etc. To this end, a feedback flow—represented by a dashed arrow—is inserted into the task net (Fig. 15). The feedback flow is refined by a data flow, along which the simulation data are propagated. Then, the simulation data are introduced into the flow diagram.

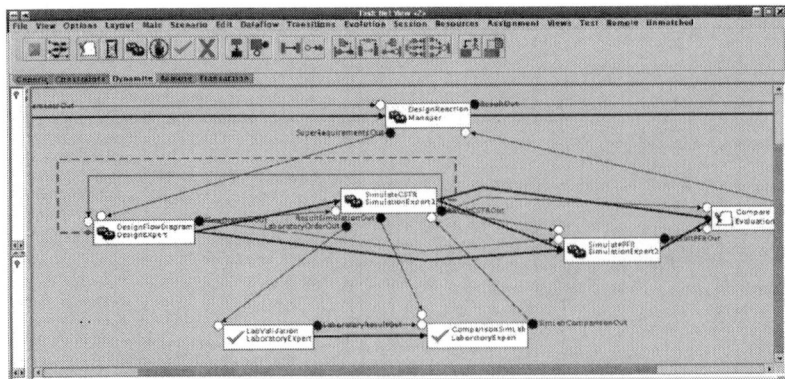

Figure 15: Feedback and simultaneous engineering (management environment).

Please note that the tasks for designing the flow diagram and for investigating the CSTR are active at the same time. The semantics of control flows is defined such that tasks connected by control flows can be active simultaneously. In this way, we support simultaneous engineering (Bullinger & Warschat, 1996). As a consequence, we cannot assume that the work context of a task is stable with respect to its inputs. Rather, a predecessor task may deliver a new version that is relevant for its successors. This is taken care of by a sophisticated release policy built into the model underlying dynamic task nets (Westfechtel, 1999).

After the alternatives CSTR and PFR have been elaborated, the evaluation expert compares all explored design alternatives. Since none of them performs satisfactorily, feedback is raised to the design task. Note this is an example of far-reaching feedback (from the end to the start of the subprocess for designing the reaction part). Here, we assume that the designer has already terminated the design task. As a consequence, the design task has to be reactivated. Reactivation is handled by creating a new task version, which may or may not be assigned to the same designer as before. New design alternatives are created, namely a CSTR–CSTR and a CSTR–PFR cascade, respectively (see again Fig. 13). Furthermore, the task net is augmented with corresponding simulation tasks (Fig. 16 [6]). After that, the new simulation tasks are delegated to simulation experts, and simulations are carried out accordingly. Eventually, the most suitable reactor alternative is selected.

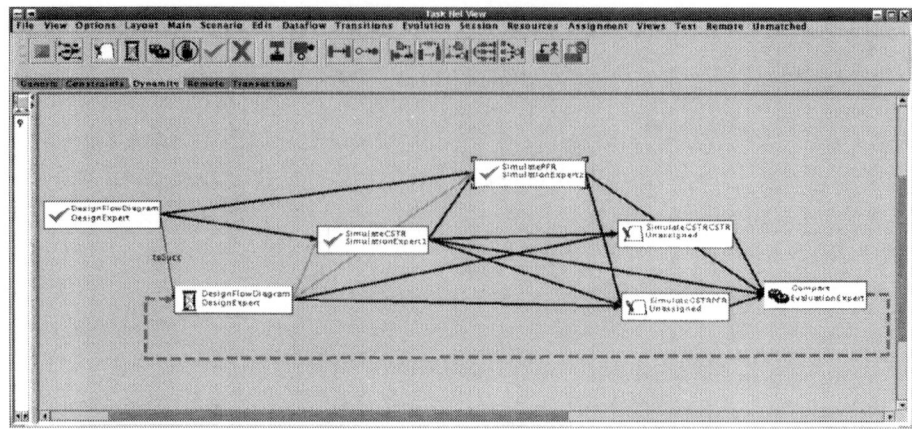

Figure 16: Far-reaching feedback (management environment).

So far, we have primarily considered the management of activities. Management of products, however, is covered as well. This is illustrated by the snapshot in Fig. 17, which is again taken from the management environment. It shows a tree view on the products of design processes on the left and a graph view on the right. Products are arranged into workspaces that are organized according to the task hierarchy. Workspaces contain sets of versioned documents.

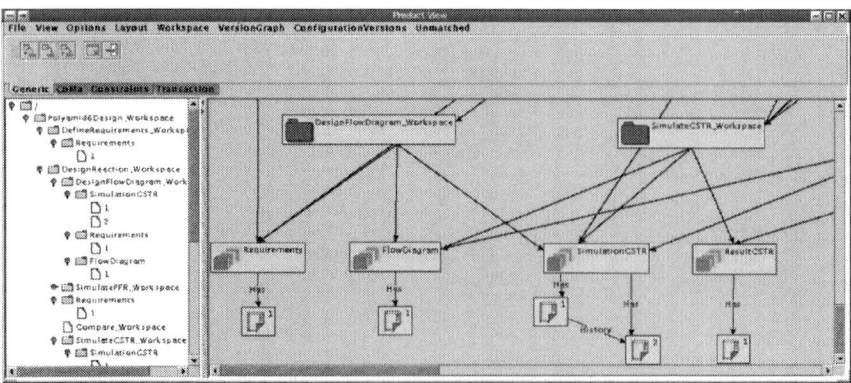

Figure 17: Product view (management environment).

Generally speaking, a version represents some state (or snapshot) of an evolving document. We distinguish between revisions, which denote temporal versions, and variants, which exist concurrently as alternative solutions to some design problem. Revisions are organized into sequences, variants result in branches. Versions are connected by history relationships. In Fig. 17, there is a history relationship between revisions 1 and 2 of Simulation CSTR, the simulation model for the CSTR reactor. In general, the version history of a document (flow diagram, simulation model, etc.) may evolve into an acyclic graph (not shown in the snapshot).

Please note that in Fig. 17 there is only one version of the flow diagram. Here, we rely on the capabilities of the flow diagram editor to represent multiple variants. Still, the flow diagram could evolve into multiple versions at the managerial level (e.g., to record snapshots at different times). Moreover, in the case of a flow diagram editor with more limited capabilities (no variants), variants would be represented at the managerial level as parallel branches in the version graph.

TECHNOLOGY TRANSFER

In this section we discuss how the obtained results can be transferred to industry in order to improve the state of the art of coordinating industrial design processes.

Ways of Technology Transfer

Let us start with discussing the different ways of technology transfer of the obtained results. Transfer ranges from evaluating the new concepts introduced above to implementing the functionality of AHEAD in an industrial context by using existing systems:

- Conceptual transfer (transfer 1): Conceptual transfer includes to see whether the concepts can be explained and are accepted, whether they solve real-world problems, and whether the extension of industrial practice can be managed. This can be achieved by demonstrating AHEAD using a prepared demo session as described in the previous section, by discussing the underlying concepts with industrial partners, etc.

- Evaluation of the demonstrator prototype (transfer 2): Here, we plan the use of the AHEAD demonstrator for a small industrial project, again in a project together with an industrial partner. This requires that control on a medium-grained level is available in the company and there is the intention to manage that level. Discussions with different industrial partners, where the AHEAD system was demonstrated, have shown that the necessity of control on medium-grained level is seen. An exemplary use of the system also includes the evaluation of the user interfaces for the work environment, the management environment, and the modeling environment.

- Realization of an industrial system with comparable functionality (transfer 3): Comparable functionality here means that the requirements stated in Section 3 and Section 4 are met, or are at least partially met. So, we aim at getting as much functionality of the AHEAD system as possible or, at least, we do not loose essential functionality. Thereby, we take existing systems and integrate them on a bottom–up basis.

Different Strategies and Common Problems

In the rest of this section we concentrate on realization transfer (transfer 3). There are two ways how an industrial system (with the functionalities of AHEAD) can look like.

The AHEAD system itself is acting as a management integration instance. Existing industrial systems are wrapped and incorporated into the integrated industrial system (Fig. 18).

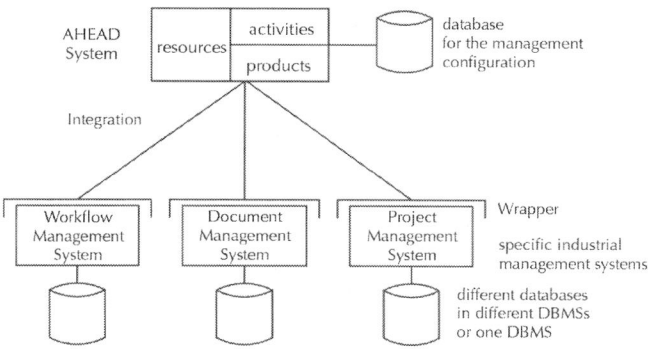

Figure 18: The AHEAD system as integrating instance of commercial systems.

We take again industrial systems for different purposes. On top of them a new system is realized which has essential AHEAD functionality. However, this system realization is using industrial components as well as aconventional industrial software development process, i.e., no rapid prototyping (Fig. 19).

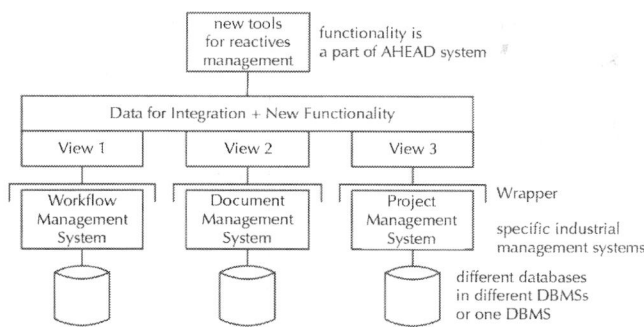

Figure 19: Realization of AHEAD functionality using industrial infrastructure and systems.

For the rest of this subsection, we give some remarks and state some implementation problems which hold for both solutions (a) and (b). The reader should note that both ways follow the overall bottom–up approachof IMPROVE as explained in Section 1 (Fig. 1) inasmuch as existing tools are re-used.

Industrial systems to be used in a bottom–up approach have to be stripped in order to concentrate on their main functionality (separation of concerns). For example, a document management system is only used for the management of documents and not for managing activities and resources, although such systems may include some restricted functionality for that purpose. Otherwise, we would have to handle activity management in different industrial systems and in the integrated system on top as well. It is difficult enough to handle it both on the level of existing systems and on the integration instance level, where different existing systems are unified and integrated.

Both variants (a) and (b) introduce wrappers for existing systems. Wrappers have two aspects. First, we need a functional interface which is coarse- (start/end) or fine-grained (invoking all commands) depending on the openness of existing systems. Second, there is a data interface in the form of a view. So, we do not access the mostly cryptic internal data of an industrial application. Instead, we define a data abstraction layer by which we get rid of data structuring details. Of course, the data views of industrial systems have to be homogeneous with respect to granularity and structure.

In both variants (a) and (b) we internally have to maintain and to retrieve the data of the management configuration. In variant (a), the management configuration is represented in the AHEAD system. In variant (b), the management configuration has to be built up by using views on existing systems. In addition, the mutual relations between these parts have to be defined (e.g., between activities and resources) in the second variant.

Discussion of Both Implementation Alternatives

We now discuss the two ways of realizing an industrial system in more detail.

Way (a) can again be split into two variants. The first variant (a1) is to have the full description for products, activities, and resources within the management configuration of AHEAD. The second variant (a2) is that the AHEAD system only contains a filtered portion of the data of the existing systems together with someintegration data. Then, the details are stored in the industrial subsystems.

In both cases (a1) and (a2) we find the semantic models of design processes in AHEAD at a central place, guaranteeing that the semantic submodels fit together. Within industrial systems a part of these models cannot be mapped (for example the integration of the three perspectives for products, activities, and resources). In solution (a1), we find the detail information for the submodels in the industrial systems as well as in AHEAD. As the access is simpler via AHEAD, we make data access and maintenance using this system. In solution (a2), we find only a part of the information in AHEAD. So, access of integrating information as well as coarse information is done via AHEAD, whereas the details are accessed using theview interfaces for industrial systems.

Let us discuss the coupling of AHEAD and the industrial systems by taking the task part of AHEAD and a workflow system as an example. As discussed in Section 3, workflow systems which allow to structure a task plan either do not regard dynamic aspects at all (static plans) or they allow subnet calling thereby distributing the static structure of a dynamically changing net over subnet definitions and the run time storage of the executing system. In solution (a1) the complete net structure is built up in AHEAD, the given workflow system is only used for executing the overall net or subnets. In solution (a2) only the global structure is kept within AHEAD whereas the detailed subnet structures are found in the workflow system. Thereby, execution is performed on a coarse level in AHEAD and on a more detailed level in the workflow system.

Let us now shortly discuss alternative (b) of above. Here, the AHEAD functionality of semantic submodels for products, activities, and resources, integration of submodels, interleaved invocation of structuring, analyzing and execution, adaptation mechanisms, etc. have to be re-implemented in the new industrial system on top of existing systems using only conventional software development techniques. This is not trivial. However, the essential features can be realized with some (remarkable) effort. Comparing both solutions (a) and (b) we can state: Way (a) is much simpler, as the AHEAD system with its functionality is a part of the integration solution. However, the AHEAD system is just a demonstrator from academia and not an industrial system. So way (b) is more likely to be used. However, then the functionality of AHEAD has to be re-implemented, a nontrivial software development task.

As a consequence of the above discussion, the question now arises why we did not realize the AHEAD system by a bottom–up approach from the very beginning. The answer to this question is that we have firstly looked for the right concepts. In order to find them, the different possibilities had to be investigated. For playing around, a realization machinery as explained in Section 4.2 is much more convenient than coding different variants. Now, as a demonstrator is available, and if it has successfully passed the different stages of proof of concept (transfer 1 and 2), we are tackling the question of how an industrial system can look like. Our research is still continuing using the implementation machinery of Section 4.2 in order to quickly find further concepts/mechanisms which, again in a second run, are transferred to industry. Therefore, also in the future, we distinguish between scientific investigation on the one hand and transfer projects on the other.

CONCLUSIONS AND OUTLOOK

We have presented the AHEAD system for managing design processes in chemical engineering. AHEAD goes beyond the functionality of commercial systems in various respects. Management is performed on a medium-grained level, allowing to effectively coordinate the work of designers. Products, activities, and resources are equally taken into account and are managed in an integrated way with the help of the management environment. Through its work environment, AHEAD is coupled with external tools supporting designers in their technical work. By seamless interleaving of planning and execution, AHEAD takes dynamic design processes into account. Finally, AHEAD may be adapted to different application domains with the help of its modeling environment.

Ongoing and future work addresses the following areas:

- Technology transfer: As discussed in Section 6, current work addresses the evolution of the AHEAD prototype into a system which may be used in an industrial context.

- Improved functionality: We are still extending the AHEAD prototype to improve its capabilities in various respects, e.g., concerning the management of distributed design processes (Becker, Jäger, Schleicher, & Westfechtel, 2001) or even more sophisticated support for process evolution (Schleicher, 2002).

- Synergy: We are integrating AHEAD more tightly with other components of the integrated design environment developed in the IMPROVE project. For example, we are designing an integration tool for coupling flow diagrams and task nets (so far, this coupling has to be performed manually), and we are investigating potentials for fine-grained process support for managers using the AHEAD system.

ACKNOWLEDGMENTS

We are indebted to all researchers who have worked in the IMPROVE project and have contributed to the AHEAD system in one way or the other. In particular, we are indebted to A. Schleicher and D. Jäger, who have made essential contributions in their Ph.D. theses, and to B. Bayer, M. Eggersmann, and W. Marquardt.

REFERENCES

1.   Ban˜ares-Alca´ntara, R. (1995). Design support systems for processengineering 1. Requirements and proposed solutions for a designprocess representation. Computers and Chemical Engineering 19(3), 267/277.

2.   Ban~ares-Alca´ntara, R., & Lababidi, H. (1995). Design support systemsfor process engineering-II. KBDS: An experimental prototype.Computers and Chemical Engineering 19 (3), 279/301.

3.   Bayer, B., Eggersmann, M., Gani, R., & Schneider, R., (2002). Casestudies in process design. In: Braunschweig & Gani, 2002.

4.   Bayer, B., Marquardt, W., Weidenhaupt, K., & Jarke, M., 2001. Aflowsheet centered architecture for conceptual design. In: Gani, R.,& Jorgensen, S. (Eds.), European symposium on computer aidedprocess engineering 11 (pp. 345/350), Elsevier.

5.   Becker, S., Haase, T., Westfechtel, B., & Wilhelms, J., 2002. Integrationtools supporting cooperative development processes in chemicalengineering. In: Ehrig, H., Kra¨mer, B.J., & Ertas, A. (Eds.),Proceedings of the sixth biennial world conference on integrateddesign and process technology (IDPT2002). Society for Design andProcess Science, Pasadena, California.

6.   Becker, S., Ja¨ger, D., Schleicher, A., & Westfechtel, B., 2001. Adelegation-based model for distributed software process management.In: Ambriola, V. (Ed.), Proceedings eighth european workshopon software process technology (EWSPT 2001). LNCS 2077(pp. 130/144). Witten, Germany: Springer.

7.   Biegler, L. T., Grossmann, I. E., & Westerberg, A. W. (1997). Systematic Methods of Chemical Process Design. Upper SaddleRiver: Prentice Hall.

8.   Blass, E. (1997). Entwicklung verfahrenstechnischer Prozesse: Methoden,Zielsuche, Lo¨sungssuche, Lo¨sungsauswahl. Berlin, Germany:Springer.

9.   Bogusch, R., Loehmann, B., & Marquardt, W. (2001). Computeraidedprocess modeling with ModKit. Computers and ChemicalEngineering 25 (7/8), 963/995.

10.   Bo¨hlen, B., Ja¨ger, D., Schleicher, A., & Westfechtel, B., 2002. UPGRADE: Building interactive tools for visual languages.

In:Callaos, N., Hernandez-Encinas, L., & Yetim, F. (Eds.), Vol. I:Information Systems Development I. Proceedings of the sixth worldmulticonference on systemics, cybernetics, and informatics (SCI2002) (pp. 17/22). Orlando, Florida.

11. Booch, G., Rumbaugh, J., & Jacobson, I. (1999). The Unified ModelingLanguage User Guide . Reading, Massachusetts: Addison Wesley.Braunschweig, B., & Gani, R. (Eds.). Software Architectures and Toolsfor Computer Aided Process Engineering. Elsevier Publishers, Vol.11.

12. Bullinger, H.-J., Warschat, J., Concurrent Simultaneous EngineeringSystems (1996). Berlin, Germany: Springer. Springer, Berlin, GermanyDerniame, J.-C., Baba, A.K., Wastell, D., Software Process: Principles,Methodology, and Technology. LNCS 1500 (1998).

13. Berlin,   Germany:Springer.Deutsches   Patentamt, 1969.   Verfahren   and   Vorrichtung   zum kontinuierlichenPolykondensieren von Lactamen. Patent number P 14 95198.5 (B78577).

14. Douglas, J. M. (1988). Conceptual Design of Chemical Processes. New York: McGraw-Hill.

15. Eggersmann, M., Hackenberg, J., Marquardt, W., & Cameron, L.,2002. Applications of modeling*/a case study from process design .In: Braunschweig & Gani, 2002.

16. Foltz, C., Killich, S., Wolf, M., Schmidt, L., & Luczak, H. (2001).Task and information modeling for cooperative work. In: M.Smith, & G. Salvendy (Eds.), Systems, Social and InternationalisationDesign Aspects of HumanComputer Interaction. Volume 2 ofthe Proceedings of HCI International 2001. Lawrence ErlbaumAssociates, Mahwah, New Jersey (pp. 172/176).

17. Georgakopoulos, D., Prinz, W., & Wolf, A.L. (Eds.), 1999. Proceedingsof the International Joint Conference on Work ActivitiesCoordination and Collaboration (WACC-99). Vol. 24/2 of ACMSIGSOFT Software Engineering Notes. San Francisco, CA: ACMPress.

18. Gerrens, H. (1981). On selection of polymerization reactors. GermanChemical Engineering 4 , 1/13.

19. Harris, S.B. (1996). Business strategy and the role of engineeringproduct data management: A literature review and summary of theemerging research questions. Proceedings of the Institution ofMechanical Engineers, Part B (Journal of Engineering Manufacture)210(B3) 207/220.

20. Jablonski, S., & Bußler, C. (1996). Workflow Management-ModelingConcepts and Architecture . Bonn, Germany: International ThomsonPublishing.

21. Ja¨ger, D., Schleicher, A., & Westfechtel, B. (1999). AHEAD: A graphbasedsystem for modeling and managing development processes.In: Nagel, et al. (pp. 325/339).

22. Ja¨ger, D., Schleicher, A., & Westfechtel, B. (1999b). Using UML forsoftware process modeling. In O Nierstrasz & M. Lemoine (Eds.),Software engineering-ESEC/FSE '99. LNCS 1687 (pp. 91/108).Toulouse, France: Springer.

23. Kerzner, H. (1998). Project Management: A Systems Approach toPlanning, Scheduling, and Controlling. New York: Wiley.

24. Kiesel, N., Schu¨rr, A., & Westfechtel, B. (1995). GRAS, a graphorientedsoftware engineering database system. Information Systems20 (1), 21/51.

25. Kohan, M., Nylon Plastics Handbook (1995). Munich, Germany: CarlHanser Verlag.

26. Krapp, C.-A., Kru¨ppel, S., Schleicher, A., & Westfechtel, B. (1998).Graph-based models for managing development processes, resources,and products. In: G. Engels & G. Rozenberg, TACT '98-6th International Workshop on Theory and Application of GraphTransformation. LNCS 1764, pp. 455/474. Paderborn, Germany:Springer.

27. Lawrence, P., Workflow Handbook (1997). Chichester, UK: Wiley.

28. Levy, S., Subrahmanian, E., Konda, S., Coyne, R., Westerberg, A.,Reich, Y., 1993. An overview of the n-dim environment.

Tech.Rep. EDRC-05-65-93, Carnegie Mellon University, Pittsburgh,Pennsylvania.

29. Lipperts S., Thißen D., 1999, CORBA wrappers for a-posteriorimanagement: An approach to integrating management withexisting heterogeneous systems. In: Proceedings third InternationalConference on Formal Methods for Open Object-Based DistributedSystems. Florence, Italy, pp. 273/280.

30. Marquardt, W., & Nagl, M., 1999. Tool integration via interfacestandardization? In: Computer Application in Process and PlantEngineering-Papers of the 36th Tutzing Symposion. Vol. 135 ofDECHEMA Monographie. Wiley VCH, Weinheim, Germany, pp.95/126.

31. McCarthy, J., & Bluestein, W., 1991. The Computing Strategy Report: Workflow's Progress. Forrester Research, Inc.

32. McGuire, M. L., & Jones, K. (1989). Maximizing the protential ofprocess engineering databases. Chemical Engineering Progress 85(11), 78/83.

33. Mostow, J. (1985). Toward better models of the design process. The AIMagazine 6 (1), 44/57.

34. Nagl, M., Schneider, R., & Westfechtel, B., Synergetische Verschra¨nkungbei der A-posteriori-Integration von Werkzeugen. In: Nagl,M., & Westfechtel, B. (Eds.), Modelle, Werkzeuge and Infrastrukturenzur Unterstu¨tzung von Entwicklungsprozessen. Wiley VCH,Weinheim, Germany, in press.

35. Nagl, M., & Westfechtel, B., Integration von Entwicklungssystemen inIngenieuranwendungen (1998). Heidelberg, Germany: Springer.

36. PIEBASE Working Group 2, 1998. Activity Model. Available fromhttp://cic.nist.gov/piebase (Accessed 8 October, 2001).

37. Pohl, K., Klamma, R., Weidenhaupt, K., Do¨mges, R., Haumer, P., &Jarke, M. (1999). Process-integrated (modelling) environments(PRIME): Foundations and implementation

framework. ACMTransactions on Software Engineering and Methodology 8 (4), 343/410.

38. Ponton, J. (1995). Process systems engineering: Halfway through thefirst century. Chemical Engineering Science 50 (24), 4045/4059.

39. Rudd, D., Powers, G., & Siirola, J. (1973). Process Synthesis. Englewood Cliffs: Prentice-Hall.

40. Schleicher, A. (1999). In: Nagl, et al., Formalizing UML-based processmodels using graph transformations. pp. 341/358.

41. Schleicher, A., 2002. Roundtrip process evolution support in a widespectrum process management system. Ph.D. thesis, AachenUniversity of Technology, Aachen, Germany.

42. Schu"ppen, A., Trossen, D., & Wallbaum, M. (2002). Shared workspacefor collaborative enginering. Annals of Cases on InformationTechnology IV, 119/130.

43. Schu"rr, A., Winter, A., & Zu"ndorf, A. (1999). The PROGRESapproach: Language and environment. In: H. Ehrig, G. Engels,H.-J. Kreowski & G. Rozen-berg. Handbook on Graph Grammarsand Computing by Graph Transformation: Applications, Languages,and Tools, vol. 2: Applications, Languages and Tools.(pp. 487/550). Singapore: World Scientific.

44. Smithers, T., & Troxell, W. (1990). Design is intelligent behaviour, butwhat's the formalism? Artificial Intelligence for Engineering Design.Analysis and Manufacturing 4 (2), 89/98.

45. Thayer, R.H. (1988). Software engineering project management: Atop/down view. In: R.H. Thayer. Tutorial: Software EngineeringProject Management, (pp. 15/54). Washington, DC: IEEE ComputerSociety Press.

46. Tichy, W.F. Configuration Management of Trends in Software , vol. 2(1994). New York: Wiley.

47. Westerberg, A. W., Subrahmanian, E., Reich, Y., & Konda, S. (1997).Designing the process design process. Computers & ChemicalEngineering 21 (S), S1/S9.

48. Westfechtel, B. (1999). Models and Tools for Managing DevelopmentProcesses. LNCS 1646. Heidelberg, Germany: Springer.

49. Whitgift, D. (1991). Methods and Tools for Software ConfigurationManagement. Wiley Series in Software Engineering Practice . NewYork: Wiley.

50. Yin, R. (1984). Case Study Research. Beverly Hills: Sage Publications.

Citations

CHAPTER 1

Omar Z. Sharaf, Mehmet F. Orhan, An overview of fuel cell technology: Fundamentals and applications, Renewable and Sustainable Energy Reviews, Volume 32, April 2014, Pages 810-853, ISSN 1364-0321, http://dx.doi.org/10.1016/j.rser.2014.01.012.

CHAPTER 2

R.J.P Williams, The fundamental nature of life as a chemical system: the part played by inorganic elements, Journal of Inorganic Biochemistry, Volume 88, Issues 3–4, February 2002, Pages 241-250, ISSN 0162-0134, http://dx.doi.org/10.1016/S0162-0134(01)00350-6.

CHAPTER 3

Y. Ngothai, M.C. Davis, Implementation and analysis of a Chemical Engineering Fundamentals Concept Inventory (CEFCI), Education for Chemical Engineers, Volume 7, Issue 1, January 2012, Pages e32-e40, ISSN 1749-7728, http://dx.doi.org/10.1016/j.ece.2011.10.001.

CHAPTER 4

Hegles Rosa de Oliveira, Alice de Oliveira de Avelar Alchorne, Fundamentals of the knowledge about chemical additives present in rubber gloves, http://dx.doi.org/10.1590/S0365-05962011000500008.

CHAPTER 5

V. B. Soares, G. L. V. Coelho, Safety study of an experimental apparatus for extraction with supercritical CO_2, http://dx.doi.org/10.1590/S0104-66322012000300023.

CHAPTER 6

Manfred Nagl, Bernhard Westfechtel, Ralph Schneider, Tool support for the management of design processes in chemical engineering, Computers & Chemical Engineering, Volume 27, Issue 2, 15 February 2003, Pages 175-197, ISSN 0098-1354, http://dx.doi.org/10.1016/S0098-1354(02)00164-3.

Index